FRP-Strengthened
Metallic Structures

FRP-Strengthened
Metallic Structures

Xiao-Ling
Zhao

CRC Press
Taylor & Francis Group
Boca Raton   London   New York

CRC Press is an imprint of the
Taylor & Francis Group, an **informa** business

CRC Press
Taylor & Francis Group
6000 Broken Sound Parkway NW, Suite 300
Boca Raton, FL 33487-2742

First issued in paperback 2017

Version Date: 20130725

ISBN 13: 978-0-415-46821-3 (hbk)
ISBN 13: 978-1-138-07433-0 (pbk)

Library of Congress Cataloging-in-Publication Data

Zhao, Xiao-Ling.
   FRP-strengthened metallic structures / Xiao-Ling Zhao.
      pages cm -- (Spon research)
   Includes bibliographical references and index.
   ISBN 978-0-415-46821-3 (hardback)
   1. Building, Iron and steel--Materials. 2. Buildings--Maintenance and repair--Materials. 3. Fiber-reinforced plastics. 4. Adhesives. I. Title.

TA684.Z4933 2014
624.1'8923--dc23
                                          2013027626

Visit the Taylor & Francis Web site at
http://www.taylorandfrancis.com

and the CRC Press Web site at
http://www.crcpress.com

Contents

Preface

A significant number of metallic structures are aging. The conventional method of repairing or strengthening aging metallic structures often involves bulky and heavy plates that are difficult to fix and prone to corrosion, as well as to their own fatigue. Fibre-reinforced polymer (FRP) has great potential for strengthening metallic structures, such as bridges, buildings, offshore platforms, pipelines, and crane structures.

The existing knowledge of the carbon fibre-reinforced polymer (CFRP)–concrete composite system may not be applicable to the CFRP–steel system because of the distinct difference between their debonding mechanisms, alongside the unique failure modes for steel members and connections. Several design and practice guides on FRP strengthening of metallic structures have been published in the United Kingdom, the United States, Italy, and Japan. However, the following topics are not covered in detail: bond behaviour between FRP and steel, strengthening of compression members, strengthening of steel tubular members, strengthening against web crippling of steel sections, and strengthening for enhanced fatigue and seismic performance. This book contains not only descriptions and explanations of basic concepts and summarises the research performed to date on the FRP strengthening of metallic structures, but also provides some design recommendations. Comprehensive, topical references appear throughout the book. It is suitable for structural engineers, researchers, and university students who are interested in the FRP strengthening technique.

This book will provide a comprehensive treatment of the behaviour and design of FRP-strengthened metallic structures, especially steel structures, based on existing worldwide research. Chapters 1 and 2 outline the applications, existing design guidance, and special characteristics of FRP composites within the context of their use in the strengthening of metallic structures. Chapter 3 deals with the bond behaviour between FRP and metal. The strengthening of members is covered in Chapter 4 (bending), Chapter 5 (compression), and Chapter 6 (bearing forces). Chapter 7 provides a description of improvement of fatigue performance.

Chapter 4 is authored by Prof. Jin-Guang Teng at Hong Kong Polytechnic University and Dr. Dilum Fernando at the University of Queensland. I am appreciative of the comments from Prof. Jin-Guang Teng on the first three chapters; Dr. Yu (Barry) Bai at Monash University on Chapters 2 and 3; Prof. Dinar Camotim at Technical University of Lisbon, Prof. Amir Fam at Queen's University, and Prof. Amr Shaat at Ain Shams University on Chapter 5; Prof. Ben Young at the University of Hong Kong on Chapter 6; and Prof. Sing-Ping Chiew at Nanyang Technological University, Singapore, and Prof. Hitoshi Nakamura at Tokyo Metropolitan University on Chapter 7. I thank Dr. Mohamed Elchalakani at Higher Colleges of Technology, Dubai Men's College, for checking the design examples in Chapters 5 and 6.

Xiao-Ling Zhao

Acknowledgments

I thank the following people for providing necessary documents, data, or photos: Dr. Mike Bambach at the University of New South Wales; Dr. Mohamed Elchalakani at Higher Colleges of Technology, Dubai Men's College; Prof. Amir Fam at Queen's University; Prof. Peng Feng at Tsinghua University; Dr. Dilum Fernando at the University of Queensland; Prof. Kent Harries at the University of Pittsburgh; Prof. Len Hollaway at Surrey University; Prof. Stuart Moy at University of Southampton; Prof. Hitoshi Nakamura at Tokyo Metropolitan University; Prof. Amr Shaat at Ain Shams University; Dr. Nuno Silvestre at Technical University of Lisbon; Sarah Witt at Fyfe Company LLC; Prof. Ben Young at the University of Hong Kong; and Dr. Zhigang Xiao, Dr. Hongbo Liu, Dr. Jimmy Haedir, and Chao Wu at Monash University.

One of the cover photographs (overhead sign structure) was reproduced by kind permission of Fyfe Company; the other (bridge pier), by kind permission of Hanshin Expressway Public Corporation, Japan. The assistance from Prof. Amir Fam at Queen's University, Sarah Witt at Fyfe Company, and Prof. Hitoshi Nakamura at Tokyo Metropolitan University to obtain the photos is greatly appreciated. I also thank Simon Bates and Tony Moore at Taylor & Francis for their advice on the book's format.

I have been so blessed to have the opportunity to collaborate with the following colleagues in the field of FRP strengthening: Prof. Riadh Al-Mahaidi at Swinburne University of Technology; Prof. Sami Rizkalla at North Carolina State University; Prof. Jin-Guang Teng at Hong Kong Polytechnic University; Prof. Raphael Grzebieta at the University of New South Wales; Prof. XiangLin Gu at Tongji University; Prof. Peng Feng at Tsinghua University; Prof. Sing-Ping Chiew at Nanyang Technological University, Singapore; Dr. Dilum Fernando at the University of Queensland; Dr. Mike Bambach at the University of New South Wales; Dr. Tao Chen at Tongji University; and Dr. Yu (Barry) Bai, Dr. WenHui Duan, and Dr. ZhiGang Xiao at Monash University. I am also grateful for the advice and encouragement from Prof. Len Hollaway at Surrey University over the past 8 years.

I gratefully acknowledge my former and current PhD students at Monash University for their contributions to part of the research presented in this book: Dr. Ahmed Al-Shawaf, Dr. Mohamed Elchalakani, Dr. Sabrina Fawzia, Dr. Jimmy Haedir, Dr. Hussein Jama, Dr. Hui Jiao, Dr. Hongbo Liu, Dr. Fidelis Mashiri, Dr. Tien Nguyen, Haider Al-Zubaidy, Tao Bai, Chao Wu, and QianQian Yu.

Finally, I thank my wife, Dan, and son, Isaac, for their continued patience and support during the preparation of this book.

Notation

The following notation is used in this book. Where nondimensional ratios are involved, both the numerator and denominator are expressed in identical units. The dimensional units for length and stress in all expressions or equations are to be taken as millimetres and megapascals (N/mm^2), respectively, unless specifically noted otherwise. When more than one meaning is assigned to a symbol, the correct one will be evident from the context in which it is used. Some symbols are not listed here because they are only used in one section and are well defined in their local context.

A_{CFRP}	Area of the CFRP composites
A_{frp}	Cross-sectional area of the FRP laminate
$A_{eff,eq}$	Effective area of the equivalent steel section
$A_{eff,flange}$	Effective width of a flange in a lipped channel section
$A_{eff,lip}$	Effective width of a lip in a lipped channel section
$A_{eff,s}$	Effective area of a lipped channel section
$A_{eff,web}$	Effective width of web in a lipped channel section
A_{es}	Cross-sectional area of an equivalent steel section
A_s	Cross-sectional area of a steel section
A_t	Cross-sectional area of the equivalent section in the Shaat and Fam stub column model
A_1	Cross-sectional area of steel beam
C_{end}	Property (C_m) at final state (after the glass transition)
$C_{initial}$	Property (C_m) at initial state (ambient temperature)
C_m	Resin-dominated material property such as bending modulus, shear modulus, or shear strength
D_t	Transformed flexural rigidity
E_1	Modulus of elasticity of an isotropic plate
E_a	Modulus of elasticity of adhesive
E_{Al}	Modulus of elasticity of aluminium
E_c	Equivalent modulus of the composites
E_{CFRP}	Modulus of elasticity of CFRP
E_f	Modulus of elasticity of CFRP fibre

E_{frp}	Modulus of elasticity of FRP
$E_{L,c,cs}$	Young's modulus of the longitudinal carbon fibres in compression
$E_{L,t,cs}$	Young's modulus of the longitudinal carbon fibres in tension
E_{steel}	Modulus of elasticity of steel
$E_{T,c,cs}$	Young's modulus of the transverse carbon fibres in compression
$E_{T,t,cs}$	Young's modulus of the transverse carbon fibres in tension
F	Crack size–dependent correction factor
F_1	Ultimate bond strength under static load
$F_{1,d}$	Ultimate bond strength under static load for a bonded joint failing in debonding
F_2	Ultimate bond strength after a preset number of fatigue cycles
F_c	Force carried by CFRP composites
F_e	Correction factor for crack shape
F_g	Correction factor for stress gradient
F_h	Correction factor for eccentricity of crack against the central axis of the plate
F_s	Correction factor for surface crack or force carried by steel plate
F_t	Correction factor for finite thickness and width of plate
F_w	SIF reduction factor considering the influence of crack length and CFRP bond width
G_a	Shear modulus of adhesive
$G_a(T)$	Shear modulus of adhesive at a certain temperature T
G_f	Interfacial fracture energy defined by the area under the bond–slip curve
I_1	Second moment of area of steel beam
I_f	Second moment of area of CFRP composites
I_s	Second moment of area of a steel section
K_n	Normal stiffness of the adhesive layer
K_s	Stress intensity factor at crack tip or shear stiffness of the adhesive layer
$K_{s,max}$	Maximum stress intensity factor
$K_{s,min}$	Minimum stress intensity factor
L	Bond length or column length
L_e	Effective bond length
$L_e(T)$	Effective bond length at a certain temperature T
L_i	Distance of the ith strain gauge from the free end of the CFRP plate
$M_{b,rd}$	Inelastic lateral buckling moment
M_{cr}	Elastic lateral buckling moment
M_p	The plastic moment capacity per unit width

M_u	Ultimate moment carrying capacity
$M_{u,frp}$	In-plane moment capacity of the FRP-plated section assuming failure by FRP rupture
$M(x)$	Moment acting on the section at x distance from the plate end
M_x	Moment of the section when neutral axis depth is x
N	Number of fatigue cycles
$N_{c,E}$	Elastic global buckling capacity
$N_{c,D}$	Distortional buckling capacity
$N_{c,L}$	Local buckling capacity
N_{cr}	Elastic critical load for torsional buckling
N_D	Critical elastic distortional column buckling load
N_L	Critical elastic local column buckling load
$N_{s,T}$	Section capacity of T-section without FRP strengthening
$N_{y,s}$	Yield capacity of a CHS
$N(x)$	Axial force acting on the section at x distance from the plate end
P	Applied tensile load
P_a	Force in the adhesive
P_c	Force in the equivalent composites
P_f	Force in the FRP fibre
P_{max}	Maximum applied load in a fatigue test
P_n	Total peeling force to be resisted by an FRP end wrap
P_{ult}	Ultimate load carrying capacity or bond strength
$P_{ult}(T)$	Ultimate load carrying capacity at a certain temperature (T)
R	Tensile strain energy of the adhesive
R_b	Web bearing capacity
R_{bb}	Web bearing buckling capacity
R_{by}	Web bearing yield capacity
T	Average total thickness of the specimen at the joint or temperature
T_g	Glass transition temperature
$V(x)$	Shear force acting on the section at x distance from the plate end
a	Fatigue crack length
a_i	Initial size of crack
a_f	Final size of crack
b	Overall flange width of a metal section
b_b	Bearing load dispersion length
b_{CFRP}	Width of CFRP
b_e	Effective width of a concrete block in steel–concrete composite section
$b_{f,I}$	Width of an I-section flange i (i = 1, 2)
b_{flange}	Flange width of a lipped channel section

b_{frp}	Width of FRP laminate
b_{lip}	Lip width of a lipped channel section
b_s	Bearing length
b_{web}	Web width of a lipped channel section
d	Overall web depth of a metal section
$d_{eff,eq}$	Effective diameter of the equivalent steel section
d_{es}	Outside diameter of an equivalent CHS section
d_f	Flange depth of an LSB section
d_s	Outside diameter of a CHS
f_{cu}	Cube compressive strength of concrete
$f_{t,a}$	Tensile strength of adhesive
f_u	Ultimate tensile strength of metal
f_y	Tensile yield stress of metal
h_c	Height of the concrete block in a steel–concrete composite section
k	Effective length factor of columns
k_e	Effective buckling length factor defined in AS 4100
k_i	Factor used to define the level of strain in the ith layer with respect to the strain in the steel
k_n	Out-of-plane stiffness of adhesive
k_t	In-plane stiffness of adhesive in the transverse direction
k_u	In-plane stiffness of adhesive in the longitudinal direction
n	Number of CFRP layers on one side of the joint
n_{GFRP}	Number of GFRP sheets
n_L	Number of longitudinal FRP layers
n_T	Number of transverse FRP layers
r	Radius of gyration
r_{ext}	External corner radius of an RHS
r_i	Inner radius of a channel section
r_{int}	Internal corner radius of an RHS
r_t	Radius of gyration of a composite section
t	Wall thickness
t_a	Thickness of the adhesive between the steel plate and the first layer of CFRP
t_{a_layer}	Thickness of one layer of adhesive
t_{CFRP}	Thickness of CFRP sheet (including CFRP sheets and adhesive)
t_{CFRP_plate}	Thickness of one layer of CFRP plate
t_{CFRP_sheet}	Thickness of one layer of CFRP fibre sheet
t_{es}	Total thickness of steel and supplanted sections
$t_{es,cs}$	Thickness due to CFRP strengthening
t_f	Flange thickness of a metal section
t_{frp}	Thickness of the FRP laminate
$t_{L,cs}$	Thickness of the longitudinal carbon fibre sheet

t_s	Wall thickness of a CHS or SHS or channel section
t_{steel}	Thickness of steel plate
$t_{T,cs}$	Thickness of the transverse carbon fibre sheet
t_w	Web thickness of a metal section
t_{we}	Equivalent web thickness
y_n	Neutral axis position
x	Neutral axis depth of an I-section (or plated I-section) from the top of the section
α	Fibre reinforcement factor
α_b	Section constant of compression members
α_c	Member slenderness reduction factor defined in AS 4100
α_g	Conversion degree of the glass transition
β_L	Modular ratio associated with the longitudinal fibres
β_T	Modular ratio related to the transverse fibres
χ	Strength reduction factor associated with global (flexural or flexural–torsional) buckling
Δ	Deformation under bearing load
ΔK_s	Range of stress intensity factor at the crack tip in a steel plate
ΔK_{th}	Threshold stress intensity factor below which fatigue crack does not propagate
$\Delta\sigma$	Stress range
δ_1	Initial slip in a bond–slip model
δ_2	Slip at the end of plateau in a bond–slip model
δ_f	Maximum slip in a bond–slip model
$\delta_{i+1/2}$	Slip at the middle point between the ith strain gauge and the $i-1$th strain gauge
ε_c	Average strain of the composites
ε_{cc}	Compressive strain at the top of the concrete block
$\varepsilon_{CFRP,c}$	Average strain of CFRP
$\varepsilon_{c,i}$	Initial compressive strain
ε_{co}	Compressive strain of concrete at the first attainment of the peak axial stress
ε_{cr}	Average flexural buckling strain
ε_{cu}	Ultimate compressive strain of concrete
ε_{frp}	Average strain of the FRP laminate
ε_{frpl}	Limiting strain of the FRP laminate
$\varepsilon_{frp,I}$	Strain at intermediate debonding strength
ε_i	Reading of the ith strain gauge counted from the free end of the CFRP plate
ε_s	Average strain of the steel plate
$\varepsilon_{s,c}$	Strain at the top surface of the top (compression) flange of an I-section
$\varepsilon_{s,i}$	Strain at a depth of h_i from the top of an I-section

$\varepsilon_{s,t}$	Strain at the bottom surface of the bottom (tension) flange of an I-section
$\varepsilon_{t,i}$	Initial tensile strain
$\varepsilon_{u,CFRP}$	Ultimate tensile strain of CFRP composites
ϕ	Capacity factor
ϕ_x	Curvature of a beam with neutral axis depth x
γ_c	Material safety factor for concrete
γ_e	Adhesive elastic shear strain
$\gamma_e(T)$	Adhesive elastic shear strain at a certain temperature (T)
γ_{frp}	Partial safety factor for FRP composites
γ_{M1}	Partial safety factor for flexure
γ_p	Adhesive plastic shear strain
$\gamma_p(T)$	Adhesive plastic shear strain at a certain temperature (T)
γ_s	Material safety factor for steel
λ_{es}	Element slenderness for an equivalent CHS
λ_n	Modified compression member slenderness
λ_s	Element slenderness
λ_T	Nondimensional slenderness
ν	Poisson's ratio of steel
ν_a	Poisson's ratio of adhesive
ρ_c	Winter reduction factor in Bambach et al. stub column model
ρ_{es}	Winter reduction factor in modified EC3 model
ρ_f	Cross-sectional area ratio in Shaat and Fam column model
σ_c	Compressive stress of concrete
$\sigma_{cr,cs}$	Elastic buckling stress of a composite plate
σ_{ese}	Elastic buckling stress
σ_{frp}	Average stress of an FRP laminate
$\sigma_{frp,u}$	Ultimate strength of an FRP laminate
σ_{max}	Maximum value of the applied cyclic stress
σ_{min}	Minimum value of the applied cyclic stress
$\sigma_n(x)$	Interfacial normal stress at x distance from the plate end
σ_o	Nominal stress in steel plate
σ_{op}	Crack-opening stress
σ_s	Average stress over the steel section
$\sigma_{s,DB}$	Average stress at the nominal cross section of the steel plate for double-sided repair
$\sigma_{s,i}$	Steel stress at a depth of h_i from the top of an I-section
$\sigma_{s,SG}$	Average stress at the nominal cross section of the steel plate for single-sided repair
$\sigma_{s,y}$	Yield stress of an I-section beam
$\sigma_{y,c}$	Yield stress of corners in a cold-formed SHS
$\sigma_{y,s}$	Yield stress of a steel section
$\sigma_{y,c,eq}$	Equivalent yield stress for corners in a cold-formed SHS

$\sigma_{y,s,eq}$	Equivalent yield stress for flat faces in a cold-formed SHS
$\tau(x)$	Interfacial shear stress at x distance from the plate end
$\tau_{i+1/2}$	Shear stress at the middle point between the ith strain gauge and the i − 1th strain gauge
τ_f	Maximum shear stress in a bond–slip model
$\tau_f(T)$	Maximum shear stress in a bond–slip model at a certain temperature (T)
ξ	Proportioning factor
AA	Aluminium Association
ACFM	Alternating current field measurement (method)
ACPD	Alternating current potential drop (method)
AISC	American Institute of Steel Construction
AISI	American Iron and Steel Institute
ASI	Australian Steel Institute
AS/NZS	Australian/New Zealand Standard
ASTM	American Society for Testing and Materials
BEM	Boundary element method
CCT	Centre-cracked tensile (steel plates)
CFRP	Carbon fibre-reinforced polymer
CHS	Circular hollow section
CIRIA	Construction Industry Research and Information Association
CSA	Canadian Standards Association
DB	Double-sided (repair)
DMTA	Dynamic mechanical thermal analysis
DSC	Differential scanning calorimetry
DSM	Direct stress method
EC3	Eurocode 3
FEM	Finite element method
FRP	Fibre-reinforced polymer
GFRP	Glass fibre-reinforced polymer
GPa	Gigapascal (kN/mm^2)
HAZ	Heat-affected zone
HM	High modulus
ICE	Institution of Civil Engineers
IIFC	International Institute for FRP in Construction
JSSC	Japan Society of Steel Construction
kN	KiloNewton
LEFM	Linear elastic fracture mechanics
LSB	LiteSteel beam
MPa	Megapascal (N/mm^2)
m	Metre
mm	Millimetre
RHS	Rectangular hollow section
SG	Single-sided (repair)

SHS	Square hollow section
SIF	Stress intensity factor
TMA	Thermomechanical analysis
TRB	Transportation Research Board
UHM	Ultra-high modulus
UV	Ultraviolet

Author

Dr. Xiao-Ling Zhao obtained his BE and ME from Shanghai JiaoTong University, China, in 1984 and 1987, respectively. He received his PhD in 1992 and his Doctor of Engineering (higher doctorate) in 2012 from the University of Sydney. He also received an MBA (executive), jointly awarded by the University of Sydney and the University of New South Wales in 2007. After two years of postdoctoral research at the University of Sydney, Dr. Zhao joined Monash University in December 1994 as an assistant lecturer. He was appointed chair of structural engineering at Monash University in November 2001.

Dr. Zhao's research interests include tubular structures and FRP strengthening of structures. He has published 7 books, 5 edited special issues in international journals, and 180 Science Citation Index journal papers. He is a member of the editorial board for three international journals. Dr. Zhao has a strong history of attracting research funds through competitive grants and industry funding with a total of $10 million, including 17 Australian Research Council Grants. He has supervised 22 PhD students to completion.

Dr. Zhao's research excellence is demonstrated by the prestigious fellowships awarded by the Royal Academy of Engineering (UK), the Swiss National Science Foundation, the Alexander von Humboldt Foundation, the Japan Society for Promotion of Science, the Chinese "1000-talent" program, the Institute of Engineers Australia's Engineering Excellence Award, and the International Institute of Welding (IIW) Thomas Medal and Kurobane Lecture award.

Dr. Zhao has chaired the International Institute of FRP for Construction (IIFC) working group on FRP-strengthened metallic structures since 2005. He has chaired the IIW (International Institute of Welding) subcommission XV-E on Tubular Structures since 2002. He chaired the Australian/New Zealand Standards Committee CS/23 from 1998 to 2002. He was elected to the Fellows of American Society of Civil Engineers (ASCE), Engineers Australia (IEAust), and International Institute of FRP for Construction (IIFC). Dr. Zhao was head of the Department of Civil Engineering at Monash University, Australia, from 2008 to 2011.

Chapter 1

Introduction

1.1 APPLICATIONS OF FRP IN STRENGTHENING METALLIC STRUCTURES

Fibre-reinforced polymer (FRP) is a composite material made of a polymer matrix reinforced with fibres. The fibres are usually glass, carbon, or aramid fibres, while the polymer is usually an epoxy, vinylester, or polyester thermosetting plastic. FRPs are commonly used in the aerospace, automotive, marine, and construction industries. FRP has a high strength-to-weight ratio and good resistance to corrosion and environmental attacks (Hollaway and Head 2001).

A large number of metallic structures in road and railway infrastructure, buildings, mining, transportation, and recreation industries are ageing. The conventional method of repairing or strengthening ageing metallic structures is to cut out and replace plating or to attach external steel plates. These plates are usually bulky, heavy, difficult to fix, and prone to corrosion, as well as to their own fatigue. FRP has a potential in strengthening structures such as bridges, buildings, offshore platforms, pipelines, and crane structures. It has been widely used to strengthen concrete structures (Neale 2000, Teng et al. 2002, Nanni 2003, Rizkalla et al. 2003, TRB 2003, Oehlers and Seracino 2004). The first field application of using FRP to strengthen metallic civil engineering structures was Tickford Bridge (Lane and Ward 2000) in the UK and Christina Creek Bridge in the United States (Miller et al. 2001). The use of such advanced material to strengthen metallic structures has become an attractive option (Cadei et al. 2004).

This book deals with FRP strengthening of metallic structures. There are two forms of FRP (plates and dry fibre sheet, as shown in Figure 1.1). The dry fibre sheet is very flexible, allowing the forming of all kinds of shapes. Typical adhesives used in FRP strengthening of metallic structures are Araldite 420 and Sikadur 30. Metallic structures include those made from cast iron, carbon steel, aluminium, cast steel, and stainless steel.

A summary of some field applications is given in Table 1.1 for cast iron structures and in Table 1.2 for steel and aluminium structures.

(a)   (b)   (c)

Figure 1.1 Picture of (a) CFRP plate, (b) dry fibre sheet, and (c) adhesive.

The applications include cast iron bridges, steel bridge girders, piers and truss members, steel beams in buildings, crane structures, pipelines, and aluminium truss-type highway overhead sign structures.

Field applications of bridge repair in the United States were presented in Miller et al. (2001) and Phares et al. (2003), and those of repairing of bridges, buildings, and submarine pipes in the UK can be found in Hollaway and Cadei (2002) and Moy (2011). Fam et al. (2006) presented a field application of FRP strengthening of damaged aluminium truss joints of highway overhead sign structures. A few recent applications in China and Japan (Zhang et al. 2011, JSCE 2012) are shown in Figures 1.2 and 1.3. The strengthening was used to increase the load carrying capacity, stiffness, or fatigue life. It should be pointed out that the application is still very limited. This is mainly due to lack of research and design guides.

1.2 IMPROVED PERFORMANCE DUE TO FRP STRENGTHENING

As shown in the field applications in Section 1.1, FRP strengthens deteriorated bridges to carry more traffic loads, adds additional flexural and torsional capacities to the corroded beams in a building to cope with increased floor loading, and repairs fatigue cracks to increase fatigue life of bridge girders and truss joints. The improved performance due to FRP strengthening was also demonstrated through various research projects, as summarised in Table 1.3 for structural flexural members, compression members, composite members, beam webs, connections under static, and fatigue loading.

1.3 CURRENT KNOWLEDGE ON FRP STRENGTHENING OF METALLIC STRUCTURES

The existing knowledge of the carbon fibre-reinforced polymer (CFRP)–concrete composite system may not be applicable to the CFRP–steel system because of the distinct difference between the debonding mechanism of the

Table 1.1 Summary of some field applications of FRP strengthening cast iron structures

Structure name, location	Year built/ strengthened	Type of structure	Type of strengthening	References
Tickford Bridge, UK	1810/1999	Oldest operational cast iron road bridge in the world	Wet lay-up CFRP (final thickness up to 10 mm)	Lane and Ward (2000)
Hythe Bridge, UK	1861/1999	Cast iron girders, brick jack arches	Four prestressed CFRP plates bonded to each of 16 girders to increase load capacity to 40-tonne vehicles	Luke (2001)
Covered ways 12 and 58, Kelso Place, London Underground, London, UK	1860/1999	Brick jack arches supported by cast iron girders	CFRP plates bonded to underside of the girders to prevent overstressing while work was carried out on the foundations of the tunnel wall	Church and Silva (2002)
Shadwell station, London Underground, London, UK	1876/2000	18 cruciform section cast iron struts in brick ventilation shaft	Up to 26 plies of ultra-high-modulus (UHM) and hollow section (HS) CFRP were used	Moy et al. (2000), Leonard (2002)
King Street Bridge, Mold, Wales, UK	1870/2000	Railway bridge consisting of six cast iron girders that sustain brick arches	Preloaded to relieve stresses and transfer some dead load to the CFRP; UHM CFRP and GFRP plates bonded to cast iron; CFRP plates tapered at ends	Farmer and Smith (2001)
Bridge EL31, Surrey Quays, London Underground, London, UK	1869/2001	Cast iron beams and columns supporting brick jack arches and early riveted through decking	HM CFRP applied to increase load capacity of bridge	Church and Silva (2002)
Maunders Road Bridge, Stoke on Trent, UK	1870/2001	Beams supporting brick jack arches, carrying road over railway line	Load-relief jacking used to transfer a proportion of dead load into CFRP and increase load capacity of bridge for heavy goods vehicles; UHM CFRP plates tapered at ends	Canning et al. (2006)
Corona Bridge in Venice, Italy	1850/2001	Pedestrian bridge consisting of three cast iron arches, with spans equal to 4 m	The arches and their decorative openings were strengthened with aramid triaxial sheets and monodirectional strips, respectively	Bastianini et al. (2004)

Table 1.2 Summary of some field applications of FRP strengthening steel and aluminium structures

Structure name, location	Year built/ strengthened	Type of structure	Type of strengthening	References
Slattocks Canal Bridge, Rochdale, UK	1936/2000	Longitudinal early riveted steel beams with reinforced concrete deck	HM CFRP plates applied to inner beams, allowing bridge to carry 40-tonne vehicles	Luke (2001)
Boots building, Nottingham, UK	1921/2001	Curved steel I-beams with corrosion	The unidirectional fibres were aligned along the direction of the beam's length for flexural strengthening; the 0–90° fibre directions were used to resist shear and torsional loading, created by the curvature of the beam	Garden (2001)
Christina Creek Bridge I-704, Newark, Delaware	1962/2001	Steel girder bridge with concrete deck on Interstate 95	Demonstration project to investigate fatigue durability and environmental resistance of the adhesive bond	Miller et al. (2001)
Bridge D65A, Acton, London, UK	1870/2002	Early riveted plate girder bridge (I-section girders with timber deck)	UHM CFRP was applied to cross girders; designed to reduce live load stresses and increase fatigue life	Moy and Bloodworth (2007)

Bridge	Year	Structure	Description	Reference
Ashland bridge, Ashland, Delaware	1860/2002	Composite bridge with steel girders and concrete deck	Strengthening with CFRP plate was conducted to reduce the stress and increase fatigue life of steel girders	Chacon et al. (2004)
Aluminium truss over Route 88 in Newark, New York	1992/2003	Aluminium truss-type highway overhead sign structures	Longitudinal FRP layers were bonded to the diagonals and wrapped around the main chord to form alternating v-patterns, followed by additional circumferential layers for anchorage	Fam et al. (2006)
7838.5S 092 Bridge, Pottawattamie County, Iowa	1938/2003	Composite bridge with steel girders and concrete deck	Strengthening with CFRP plate was conducted to reduce the stress and increase fatigue life of steel girders	Phares et al. (2003)
3903.0S 141 Bridge, Guthrie, Iowa	1955/2003	Highway bridge with fabricated I-section girders	CFRP rods were used as tendons; CFRP rods are anchored on the bottom side of the web, and prestress was then applied to them to reduce the stress of the steel girders	Phares et al. (2003)
I-10 Bridge, Las Cruces, New Mexico	1957/2008	Highway bridge with steel I-beam and cross-bracing diaphragms	The repair was conducted with laminated patches of Boron fibres, which were hardened by being heated and cured at a temperature of 107°C for 3 h on the construction site	Roach et al. (2008)

Source: Adapted from Fernando, D., Yu, T., Teng J.G., and Zhao X.L., *Thin-Walled Structures*, 47(10), 1020–1028, 2009.

(a)

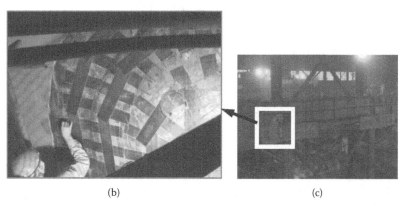

(b)   (c)

Figure 1.2 CFRP repairing of oil pipelines and crane girders in China. (a) Repair of an oil pipeline, Daqing, China. (Courtesy of Prof. P. Feng, Tsinghua University, China.) (b) Close-up view. (c) Repair of a steel crane girder. (Courtesy of Dr. Y. X. Yang, MCC, China.)

former and latter, alongside the unique failure modes for steel members and connections. The debonding in the CFRP-concrete system is mainly caused by concrete fracture, whereas the weakest link in the CFRP–steel system is the adhesive. The Young's modulus of steel is about six times that of concrete. The composite action between CFRP and steel would be different than that between CFRP and concrete. The effect of environmental conditions (extreme hot or cold weather, moisture, thermal cycling, and salt water spray) on debonding failure is different for these two systems. Concrete tends to creep and shrink, which are not characteristics of steel. The incompatibility of thermal coefficients for CFRP and concrete may

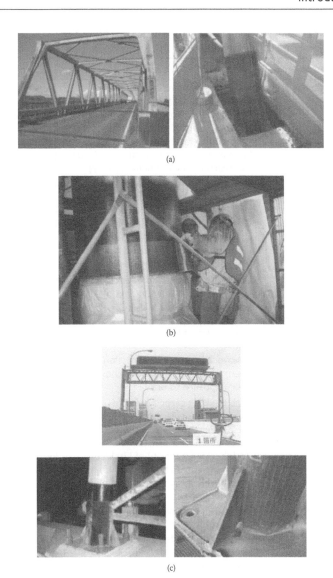

Figure 1.3 CFRP repairing of bridges in Japan. (Courtesy of Prof. H. Nakamura, Tokyo Metropolitan University, Japan.) (a) Honjo bridge, Japan. (From JSCE, Advanced Technology of Repair and Strengthening of Steel Structures Using Externally-Bonded FRP Composites, Hybrid Structure Reports 65, Japan Society of Civil Engineers, Tokyo, 2012.) (b) Hanshin bridge, Japan. (From Tokubayashi, M., and Hokazono, M., *Design and Construction for Seismic Retrofit of Steel Bridge Piers Using Carbon Fiber Sheets*, technical report, 21, 44–51, 2003.) (c) Nagoya expressway, Japan (From JSCE, Advanced Technology of Repair and Strengthening of Steel Structures Using Externally-Bonded FRP Composites, Hybrid Structure Reports 05, Japan Society of Civil Engineers, Tokyo, 2012.)

Table 1.3 Some examples of improved performance due to FRP strengthening in several research projects

Area of strengthening	Improved performance	Reference
Flexural members	• Increased moment capacity and ductility in general • For circular hollow sections, slender sections (which cannot reach first yield due to local buckling) could become noncompact sections (which can reach plastic moment capacity)	Haedir et al. (2009), Photiou et al. (2006), Seica and Packer (2007), Schnerch et al. (2007)
Compression members	• Compression strength increases about 20% for steel hollow sections and 70% for thin cylindrical shells • Compression strength increases about 15% for channel sections • Increased energy absorption under large deformation compression force	Shaat and Fam (2006), Silvestre et al. (2008), Bambach and Elchalakani (2007), Teng and Hu (2007)
Composite members	• The static strength increased by 55% and 140% when the numbers of CFRP layers were 2 and 4, respectively • Increased ductility to resist cyclic loading • Recovered certain compression strength of concrete-filled tubes after exposure to fire	Xiao et al. (2005), Tao et al. (2007a, 2007b), Teng et al. (2007)
Beam web	• Web buckling capacity of cold-formed rectangular hollow sections (RHSs) increases about 1.5 to 2.5 times • Web buckling capacity of LiteSteel beams increases about 3 to 5 times • Web buckling capacity of aluminium RHS increases up to 3.5 times	Zhao et al. (2006), Zhao and Al-Mahaidi (2009), Wu et al. (2012), Fernando et al. (2009)
Connections under static loading	• Recovered the full yield capacity, which was lost 50% due to HAZ softening, of welded very high-strength steel tubes • Recovered the full connection capacity of aluminium K-joints if CFRP is used or about 80% connection capacity if GFRP is used	Jiao and Zhao (2004), Fam et al. (2006)
Plates or connections under fatigue loading	• Fatigue life of plates increases 3 to 8 times • Fatigue life of welded web gusset joint increases 4 to 10 times • Fatigue life of cross-beam connection extended twice as the original fatigue life	Colombi et al. (2003), Liu et al. (2009), Nakamura et al. (2009), Xiao and Zhao (2012)

cause significant stresses to develop at the bond line during large swings in temperature. The difference in thermal coefficients is even larger between steel and CFRP. There is a potential galvanic corrosion problem associated with strengthening of steel members using CFRP. Some failure modes are unique for steel members and connections, such as local buckling, overall buckling, yielding, post-yield membrane action, heat-affected zone (HAZ) softening after welding, and fatigue crack propagation.

The first research study on FRP strengthening of metallic structures in the civil engineering field was probably carried out by Sen and Liby (1994). Extensive research on FRP strengthening of metallic structures was conducted in the last 10 years, mainly led by members of the International Institute for FRP in Construction (IIFC). Several IIFC international conferences on FRP in construction were held (Teng et al. 2001, Seracino 2004, Mirmiran and Nanni 2006, Motavalli 2008, Ye et al. 2010, Monti 2012, Smith 2007, Sim 2009, Ueda 2012) where many papers on FRP strengthening of metallic structures were presented. There were about 400 papers published in major journals and conferences from 2001 to 2012 on this topic. Several state-of-the-art papers were also published recently on FRP-strengthened metallic structures, such as Hollaway and Cadei (2002), Shaat et al. (2004), Zhao and Zhang (2007), and Teng et al. (2012).

Hollaway and Cadei (2002) had in-depth coverage of the following aspects:

1. In-service problems associated with advanced polymer composite and metallic adherends
2. Adhesive bonding in terms of surface preparation and durability
3. Durability of FRP composites in the civil environment
4. Prestressing FRP plates before bonding to metallic beams
5. Field applications

Shaat et al. (2004) addressed the following issues in detail:

1. Retrofit of steel girders
2. Fatigue life improvement
3. Surface preparation and other means to avoid debonding
4. Durability of steel members retrofitted with FRP
5. Field applications

Zhao and Zhang (2007) complemented the above documents by reviewing the following topics: bond between steel and FRP, strengthening of steel hollow section members, and fatigue crack propagation in the FRP-steel system. Teng et al. (2012) provided a critical review on steel surface preparation for adhesive bonding, selection of a suitable adhesive, bond behaviour between FRP and steel and its appropriate modelling, and flexural strengthening of steel beams and concrete-filled steel tubes through external FRP confinement.

Several design and practice guides on FRP strengthening of metallic structures were published in the UK, United States, Italy, and Japan. The first one in the UK was the *ICE Design and Practice Guide* written by Moy (2001), which was replaced by CIRIA C595 (Cadei et al. 2004). The CIRIA Design Guide provided detailed design guidance for strengthening metallic structures using externally bonded FRP, in particular:

1. Materials (metallic materials including cast iron and wrought iron, and FRP strengthening materials)
2. Conceptual design
3. Structural behaviour and analysis, especially I-sections with FRP
4. Design detailing the reduction of the stress concentration at the end of a plate
5. Installation and quality control
6. Operation (inspection and maintenance, owners' responsibilities)

Schnerch et al. (2007) in the United States published a design guideline mainly focusing on the use of externally bonded high-modulus carbon fibre-reinforced polymer (HM CFRP) materials to strengthen steel concrete composite bridges. The highlight of the design guidelines can be summarised as the following:

1. Installation techniques, including surface preparation of the materials, the application of the adhesive, and the detailing of the strengthening
2. Structural design criteria for the use of high-modulus CFRP materials as a flexural strengthening system of typical steel–concrete composite bridge girders
3. A bond model to calculate the shear and peel stresses within the adhesive thickness
4. A worked example to illustrate the proposed flexural design approach

The National Research Council (NRC) in Italy published a design guide in 2007 (NRC 2007) that was based on the UK and U.S. design guides. The Transportation Research Board (TRB) Structural FRP Subcommittee AFF80 is preparing a state-of-the-art document to deal with FRP strengthening of steel bridge girders.

The Japan Society of Civil Engineers published a report entitled *Advanced Technology of Repair and Strengthening of Steel Structures Using Externally-Bonded FRP Composites* (JSCE 2012). The highlight of the report is a detailed summary of field applications of FRP composites in repairing bridges.

There are a few books on the subject of FRP composites in construction (Teng et al. 2002, Hollaway and Head 2001, Oehlers and Seracino 2004, Bank 2006), but none of them deal with the subject of FRP strengthening of metallic structures.

The following topics are not covered in any detail in the above-mentioned guidelines: bond behaviour between FRP and steel, strengthening of compression members, strengthening of steel tubular members, strengthening against web crippling of steel sections, and strengthening for enhanced fatigue and seismic performance. The present book not only contains descriptions and explanations of basic concepts and summarises the research performed to date on the FRP strengthening of metallic structures, but also provides some design recommendations. Up-to-date references are given throughout the book.

1.4 LAYOUT OF THE BOOK

This book will provide a comprehensive treatment of the behaviour and design of FRP-strengthened metallic structures based on existing worldwide research. The scope of the book is indicated by the overview given below. Chapters 1 and 2 outline the applications, existing design guidance, and special characteristics of FRP composites within the context of their use in the strengthening of metallic structures. Chapter 3 deals with the bond behaviour between FRP and metal. The strengthening of members is covered in Chapter 4 (bending), Chapter 5 (compression), and Chapter 6 (bearing forces). Improvement of fatigue performance is described in Chapter 7. Future work is pointed out at the end of each chapter. While the strengthening of steel structures is the major focus of the book, the strengthening of other metallic structures, such as aluminium structures, is also given appropriate attention.

REFERENCES

Bambach, M.R., and Elchalakani, M. 2007. Plastic mechanism analysis of steel SHS strengthened with CFRP under large axial deformation. *Thin-Walled Structures*, 45(2), 159–170.

Bank, L. 2006. *Composites for construction: Structural design with FRP materials.* Hoboken, NJ: John Wiley & Sons.

Bastianini, F., Ceriolo, L., Di Tommaso, A., and Zaffaroni, G. 2004. Mechanical and nondestructive testing to verify the effectiveness of composite strengthening on historical cast iron bridge in Venice, Italy. *Journal of Materials in Civil Engineering*, 16(5), 407–413.

Cadei, J.M.C., Stratford, T.J., Hollaway, L.C., and Duckett, W.H. 2004. *C595—Strengthening metallic structures using externally bonded fibre-reinforced composites.* London: CIRIA.

Canning, L., Farmer, N., Luke, S., and Smith, I. 2006. Recent developments in strengthening technology and the strengthening/reconstruction decision. Presented at Railway Bridges Today and Tomorrow, Network Rail, Marriott Hotel City Centre, Bristol, November 22–23.

Chacon, A., Chajes, M., Swinehart, M., Richardson, D., and Wenczel, G. 2004. Application of advanced composites to steel bridges: A case study on the Ashland bridge (Delaware-USA). Presented at Proceedings of Advanced Composite Materials in Bridges and Structures, Calgary, July 20–23.

Church, D.G., and Silva, T.M.D. 2002. Application of carbon fibre composites at covered ways 12 and 58 and bridge EL. In *Proceedings of Inaugural International Conference on the Use of Advanced Composites in Construction*, Thomas Telford, London, pp. 491–500.

Colombi, P., Bassetti, A., and Nussbaumer, A. 2003. Delamination effects on cracked steel members reinforced by prestressed composite patch. *Theoretical and Applied Fracture Mechanics*, 39(1), 61–71.

Fam, A., Witt, S., and Rizkalla, S. 2006. Repair of damaged aluminum truss joints of highway overhead sign structures using FRP. *Construction and Building Materials*, 20(10), 948–956.

Farmer, N., and Smith, I. 2001. King street railway bridge strengthening of cast iron girders with FRP composites. Presented at Proceedings of the 9th International Conference on Structural Faults and Repairs, London, July 4–6.

Fernando, D., Yu, T., Teng, J.G., and Zhao, X.L. 2009. CFRP strengthening of rectangular steel tubes subjected to end bearing loads: Effect of adhesive properties and finite element modelling. *Thin-Walled Structures*, 47(10), 1020–1028.

Garden, H.N. 2001. Use of composites in civil engineering infrastructure. *Reinforced Plastics*, 45(7–8), 44–50.

Haedir, J., Bambach, M.R., Zhao, X.L., and Grzebieta, R.H. 2009. Strength of circular hollow sections (CHS) tubular beams externally reinforced by carbon FRP sheets in pure bending. *Thin-Walled Structures*, 47(10), 1136–1147.

Hollaway, L.C., and Cadei, J. 2002. Progress in the technique of upgrading metallic structures with advanced polymer composites. *Progress in Structural Engineering and Materials*, 4(2), 131–148.

Hollaway, L.C., and Head, P.R. 2001. *Advanced polymer composites and polymers in the civil infrastructure*. Oxford: Elsevier.

Jiao, H., and Zhao, X.L. 2004. CFRP strengthened butt-welded very high strength (VHS) circular steel tubes. *Thin-Walled Structures*, 42(7), 963–978.

JSCE. 2012. *Advanced technology of repair and strengthening of steel structures using externally-bonded FRP composites* (in Japanese). Hybrid Structure Reports 05. Tokyo: Japan Society of Civil Engineers.

Lane, I.R., and Ward, J.A. 2000. *Restoring Britain's bridge heritage*. Institution of Civil Engineers (South Wales Association) Transport Engineering Group Award.

Leonard, A.R. 2002. The design of carbon fibre composite (CFC) strengthening for cast iron struts at Shadwell station vent shaft. In *Proceedings of International Conference on Composites in Construction (ACIC 2002)*, Southampton University, UK, April 15–17, ed. Shenoi, R.A., Moy, S.S.J., and Hollaway, L.C. London: Thomas Telford.

Liu, H.B., Al-Mahaidi, R., and Zhao, X.L. 2009. Experimental study of fatigue crack growth behaviour in adhesively reinforced steel structures. *Composites Structures*, 90(1), 12–20.

Luke, S. 2001. The use of carbon fibre plates for the strengthening of two metallic bridges of an historic nature in the UK. In *Proceedings of FRP Composites in Civil Engineering (CICE 2001)*, Hong Kong, December, 12–15, ed. Teng, J.G., pp. 975–983. Oxford: Elsevier.

Miller, T.C., Chajes, M.J., Mertz, D.R., and Hastings, J.N. 2001. Strengthening of a steel bridge girder using CFRP plates. *Journal of Bridge Engineering*, 6(6), 514–522.

Mirmiran, A., and Nanni, A. 2006. *Proceedings of the Third International Conference on FRP Composites in Civil Engineering (CICE 2006)*, Miami, December 13–15. Miami: Florida International University.

Monti, G. 2012. *Proceedings of the Sixth International Conference on FRP Composites in Civil Engineering (CICE 2012)*, Rome, June 13–15. Rome: Sapienza Università di Roma.

Motavalli, M. 2008. *Proceedings of the Fourth International Conference on FRP Composites in Civil Engineering (CICE 2008)*, Zurich, July 22–24. Zurich: EMPA (Swiss Federal Laboratories for Materials Science and Technology).

Moy, S. 2001. *ICE design and practice guides—FRP composites life extension and strengthening of metallic structures*. London: Thomas Telford Publishing.

Moy, S. 2011. Case studies: FRP repair of steel and cast iron structures in Great Britain. *FRP International*, 8(3), 5–8.

Moy, S.S.J., Barnes, F., Moriarty, J., Dier, A.F., Kenchington, A., and Iverson, B. 2000. Structural upgrading and life extension of cast iron struts using carbon fibre reinforced composites. In *Proceedings of the 8th International Conference on Fibre Reinforced Composites*, Newcastle, UK, September 2000, pp. 3–10.

Moy, S.S.J., and Bloodworth, A.G. 2007. Strengthening a steel bridge with CFRP composites. *Proceedings of the Institution of Civil Engineers, Structures and Buildings*, 160(SB2), 81–93.

Nakamura, H., Jiang, W., Suzuki, H., Maeda, K.I., and Irube, T. 2009. Experimental study on repair of fatigue cracks at welded web gusset joint using CFRP strips. *Thin-Walled Structures*, 47(10), 1059–1068.

Nanni, A. 2003. North American design guidelines for concrete reinforcement and strengthening using FRP: Principles, applications and unresolved issues. *Construction and Building Materials*, 17(6–7), 439–446.

National Research Council. 2007. *Guidelines for the design and construction of externally bonded FRP systems for strengthening existing structures—Metallic structures (preliminary study)*. CNR-DT 202/2005. Rome, Italy: National Research Council Advisory Committee on Technical Recommendations for Construction.

Neale, K.W. 2000. FRPs for structural rehabilitation: A survey of recent progress. *Progress in Structural Engineering and Materials*, 2(2), 133–138.

Oehlers, D.J., and Seracino, R. 2004. *Design of FRP and steel plated RC structures: Retrofitting beams and slabs for strength, stiffness and ductility*. Oxford: Elsevier.

Phares, B.M., Wipf, T.J., Klaiber, F.W., Abu-Hawash, A., and Lee, Y.S. 2003. Strengthening of steel girder bridges using FRP. Presented at Proceedings of the 2003 Mid-Continent Transportation Research Symposium, Ames, IA, August 21–22.

Photiou, N.K., Hollaway, L.C., and Chryssanthopoulos, M.K. 2006. Strengthening of an artificially degraded steel beam utilizing a carbon/glass composite system. *Construction and Building Materials*, 20(1–2), 11–21.

Rizkalla, S., Tarek, H., and Hassan, N. 2003. Design recommendations for the use of FRP for reinforcement and strengthening of concrete structures. *Progress in Structural Engineering and Materials*, 5(1), 16–28.

Roach, D., Rackow, K., Delong, W., and Franks, E. 2008. In-situ repair of steel bridges using advanced composite materials. In *Proceedings of Tenth International Conference on Bridge and Structure Management*, Buffalo, NY, October 20–22, 2008, pp. 269–285. Transportation Research Circular E-C128.

Schnerch, D., Dawood, M., Rizkalla, S., and Sumner, E. 2007. Proposed design guidelines for strengthening steel bridges with FRP materials. *Construction and Building Materials*, 21(5), 1001–1010.

Seica, M.V., and Packer, J.A. 2007. FRP materials for the rehabilitation of tubular steel structures, for underwater applications. *Composite Structures*, 80(3), 440–450.

Sen, R., and Liby, L. 1994, August. *Repair of steel composite bridge sections using CFRP laminates*. Final report. Florida and U.S. Department of Transportation. Springfield, VA: National Technical Information Service.

Seracino, R. 2004. *Proceedings of the Second International Conference on FRP Composites in Civil Engineering (CICE 2004)*, Adelaide, December 8–10. London: Taylor & Francis.

Shaat, A., and Fam, A. 2006. Axial loading tests on CFRP-retrofitted short and long HSS steel columns. *Canadian Journal of Civil Engineering*, 33(4), 458–470.

Shaat, A., Schnerch, D., Fam, A., and Rizkalla, S. 2004. Retrofit of steel structures using fiber-reinforced polymers (FRP): State-of-the-art. Presented at Transportation Research Board (TRB) Annual Meeting, Washington, DC, January 11–15.

Silvestre, N., Young, B., and Camotim, D. 2008. Non-linear behaviour and load-carrying capacity of CFRP-strengthened lipped channel steel columns. *Engineering Structures*, 30(10), 2613–2630.

Sim, J. 2009. *Proceedings of the Second Asia-Pacific Conference on FRP in Structures (APFIS 2009)*, Seoul, Korea, December 9–11. Seoul: Hanyang University.

Smith, S.T. 2007. *Proceedings of the First Asia-Pacific Conference on FRP in Structures (APFIS 2007)*, Hong Kong, December 12–14. Hong Kong: University of Hong Kong.

Tao, Z., Han, L.H., and Zhuang, J.P. 2007a. Axial loading behaviour of CFRP strengthened concrete-filled steel tubular sub columns. *Advances in Structural Engineering—An International Journal*, 10(1), 37–46.

Tao, Z., Han, L.H., and Wang, L.L. 2007b. Compressive and flexural behaviour of CFRP-repaired concrete-filled steel tubes after exposure to fire. *Journal of Constructional Steel Research*, 63(8), 1116–1126.

Teng, J.G. 2001. *Proceedings of the First International Conference on FRP Composites in Civil Engineering (CICE 2001)*, Hong Kong, December 12–15. Oxford: Elsevier.

Teng, J.G., Chen, J.F., Smith, S.T., and Lam, L. 2002. *FRP—Strengthened RC structures*. West Sussex: John Wiley & Sons.

Teng, J.G., and Hu, Y.M. 2007. Behaviour of FRP-jacketed circular steel tubes and cylindrical shells under axial compression. *Construction and Building Materials*, 21(4), 827–838.

Teng, J.G., Yu, T., and Fernando, D. 2012. Strengthening of steel structures with fiber-reinforced polymer composites. *Journal of Constructional Steel Research*, 78, 131–143.

Teng, J.G., Yu, T., Wong, Y.L., and Dong, S.L. 2007. Hybrid FRP–concrete–steel tubular columns: Concept and behaviour. *Construction and Building Materials*, 21(4), 846–854.

Tokubayashi, M., and Hokazono, M. 2003. *Design and construction for seismic retrofit of steel bridge piers using carbon fiber sheets.* Technical report, vol. 21, pp. 44–51. Hanshin Expressway Public Corporation.

TRB. 2003. *Application of fibre reinformed polymer composites to the highway infrastructure.* NCHRP (National Cooperative Highway Research Program) Report 503. Washington, DC: Transportation Research Board.

Ueda, T. 2012. *Proceedings of the Third Asia-Pacific Conference on FRP in Structures (APFIS 2012)*, Sapporo, February 2–4. Sapporo: Hokkaido University.

Wu, C., Zhao, X.L., Duan, W.H., and Phipat, P. 2012. Improved end bearing capacities of sharp corner aluminum tubular sections with CFRP strengthening. *International Journal of Structural Stability and Dynamics*, 12(1), 109–130.

Xiao, Y., He, W.H., and Choi, K.K. 2005. Confined concrete-filled tubular columns. *Journal of Structural Engineering*, 131(3), 488–497.

Xiao, Z.G., and Zhao, X.L. 2012. CFRP repaired welded thin-walled cross-beam connections subject to in-plane fatigue loading. *International Journal of Structural Stability and Dynamics*, 12(1), 195–211.

Ye, L.P., Feng, P., and Yue, Q.R. 2010. *Proceedings of the Fifth International Conference on FRP Composites in Civil Engineering (CICE 2010)*, Beijing, September 27–29. Beijing: Tsinghua University.

Zhang, H.T., Xue, X.X., Zheng, Y., and Feng, P. 2011. Using CFRP to repair the steel pipe with fatigue cracks. *Advanced Materials Research*, 146–147, 1086–1089.

Zhao, X.L., and Al-Mahaidi, R. 2009. Web buckling of Lite-Steel beams strengthened with CFRP subjected to end bearing forces. *Thin-Walled Structures*, 47(10), 1029–1036.

Zhao, X.L., Fernando, D., and Al-Mahaidi, R. 2006. CFRP strengthened RHS subjected to transverse end bearing force. *Engineering Structures*, 28(11), 1555–1565.

Zhao, X.L., and Zhang, L. 2007. State of the art review on FRP strengthened steel structures. *Engineering Structures*, 29(8), 1808–1823.

Chapter 2

FRP composites and metals

2.1 GENERAL

Fibre-reinforced polymer (FRP) composites consist of a fibre matrix made from materials such as carbon, glass, or aramid fibres embedded in a resin matrix (Hollaway and Spencer 2000). The fibres of the composite matrix can be positioned into different orientations to most efficiently follow the stress distributions of the structure. An FRP sheet is flexible enough to strengthen curved surfaces. FRP materials are also resistant to corrosion, and hence maintenance and painting can be kept to a minimum (Tavakkolizadeh and Saadatmanesh 2001). As mentioned in Chapter 1, the weakest link in the FRP-metal system is the adhesive. The bond behaviour between FRP and metal depends on the material properties of adhesives. This chapter deals with the properties of FRP, adhesives, and metals.

2.2 FIBRE-REINFORCED POLYMER

For strengthening metallic structures carbon (CFRP) and glass (GFRP) fibre-reinforced polymers are commonly used. FRP is available in two types: FRP plate, as shown in Figure 1.1(a), and FRP dry fibre sheet, as shown in Figure 1.1(b). FRP dry fibre sheet is bonded to a metal surface via a wet lay-up process where epoxy is applied to form an FRP sheet. The FRP sheet could be used to strengthen curved surfaces. FRP sheets and pultruded FRP plates are called FRP laminates in this book. The thickness of FRP plates is about 1.5 mm, whereas the thickness of FRP dry fibre sheet is about 0.15 mm. The important parameters to define the properties of FRP laminates are modulus of elasticity, ultimate tensile strength, and ultimate tensile strain. Typical mechanical property values of FRP laminates can be found in Hollaway (2010).

2.2.1 Carbon fibre-reinforced polymers

CFRP laminates are suitable for the strengthening of steel structures because their modulus of elasticity is about the same or higher than that of steel. CFRPs have been classified as different categories (e.g., normal-modulus CFRP, high-modulus CFRP, and ultra-high-modulus CFRP) in literature according to their modulus of elasticity (E_{CFRP}). The modulus (E_{CFRP}) of commonly used CFRP sheets (rather than the dry fibres) varies from 230 to 640 GPa, whereas the E_{CFRP} of commonly used CFRP plates (rather than the raw carbon fibres) varies from 150 to 450 GPa. For example, CFRPs with an E_{CFRP} in the lower end are often called normal-modulus CFRPs, whereas CFRPs with E_{CFRP} in the upper end are often called high modulus by some researchers (e.g., Schnerch et al. 2007, Fawzia et al. 2005, 2007). The terms *high-modulus CFRP* and *ultra-high-modulus CFRP* are used by others (e.g., Hollaway and Cadei 2002, Photiou et al. 2006) to describe the above two categories. In this book the terms *normal modulus* and *high modulus* are adopted. The normal-modulus CFRP has a modulus of elasticity up to 250 GPa, whereas the high-modulus CFRP has a modulus of elasticity above 250 GPa (Shaat et al. 2004, Zhao and Zhang 2007, Hollaway 2010). The normal-modulus CFRP usually has a higher tensile strength (e.g., about 2500 MPa) than that (e.g., about 1200 MPa) of high-modulus CFRP. These properties are governed by the strength and stiffness of the component parts, the method of manufacture, and the orientation and density of the fibres (Hollaway and Spencer 2000).

The measured values of typical CFRP plate and CFRP sheet are listed in Table 2.1. Typical stress–strain curves are shown in Figure 2.1.

Table 2.1 Typical values of measured material properties of CFRP sheets and plates in tension

CFRP	Types	Modulus of elasticity (GPa)	Ultimate tensile strength (MPa)	Ultimate tensile strain (%)	References
CFRP sheets	High modulus	552	1175	0.20	Fawzia (2007)
	Normal modulus	230	2675	1.20	
CFRP plates	High modulus	479	1607	0.36	Wu et al. (2010)
		338	1186	Not reported	Lanier et al. (2009)
	Normal modulus	156	2691	1.72	Seracino et al. (2007)
		171	2830	1.55	Sena Cruz et al. (2006)

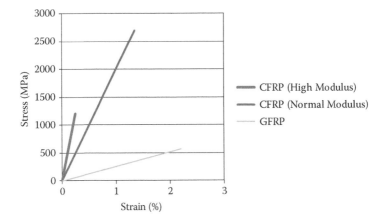

Figure 2.1 Stress–strain curves of FRP (schematic view).

2.2.2 Glass fibre-reinforced polymers

A GFRP sheet has lower stiffness and strength values than CFRP, with tensile modulus values around 30 GPa and ultimate tensile strengths ranging from 200 to 500 MPa (Fam et al. 2006, Pantelides et al. 2003). However, these materials are still used in strengthening metallic structures such as aluminium. They are particularly effective in preventing galvanic corrosion if placed in between the metal and CFRP (Tavakkolizadeh and Saadatmanesh 2001). Furthermore, they are cheaper than CFRP composites. Many different grades of glass are used. The most commonly used one in the construction industry is E-glass fibre (Cadei et al. 2004). These fibres have low alkali content and also have good resistance to heat and electricity.

2.3 ADHESIVES

A variety of adhesives can be used to bond composite materials to metallic structures. The most widely used are epoxy adhesives (Hollaway and Cadei 2002). The use of adhesives depends on the curing conditions and must be compatible with the adherends. Due to the fact that the bond is the weakest component of an FRP-steel system, it is important to understand properties such as modulus of elasticity, ultimate tensile strength, and ultimate tensile strain. Table 2.2 lists typical values of measured material properties of several types of adhesives used to bond steel. Measured stress-stain curves of five types of adhesives are plotted in Figure 2.2.

Table 2.2 Typical values of measured material properties of adhesives in tension

Adhesive	Modulus of elasticity (MPa)	Ultimate tensile strength (MPa)	Ultimate tensile strain (%)	Reference
Araldite 2015	1750	14.7	1.51	Fernando et al. (2009)
Araldite 420	1828	21.5	2.89	
FIFE-Tyfo	3975	40.7	1.11	
Sikadur 30	11,250	22.3	0.30	
Sikadur 330	4820	31.3	0.75	
Araldite 420	1901	28.6	2.40	Fawzia (2007)
Sikadur 30	9282	24.0	0.30	

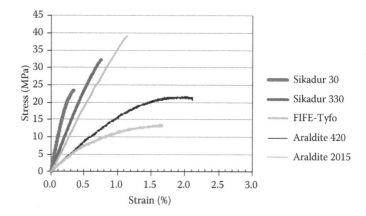

Figure 2.2 Stress–strain curves of adhesives. (Adapted from Fernando, D. et al., *Thin-Walled Structures*, 47(10), 1020–1028, 2009.)

Adhesives usually consist of two parts: a resin and a polymerising agent (Hollaway et al. 2006). Adhesives take several days to cure under ambient temperature. Post-curing can be performed at elevated temperatures, ideally to at least 50°C, to reach 100% polymerisation of the adhesive. The post-cure temperature should be carefully considered because of the glass transition temperature (T_g). When temperature is above T_g, adhesives will become soft, which leads to bond strength reduction. It was suggested by Hollaway and Cadei (2002) that adhesives have a T_g of 30°C higher than the service temperature. This will also minimise creep effects, which can arise over time if temperatures in the field are too high.

Different types of adhesives (epoxy, acrylic, and methacrylates) were used by El Damatty and Abushagur (2003) to perform a shear lap test on hollow steel sections to determine which adhesive had the highest bond strength where GFRP was used. It was found that the methacrylate (MA) 420 adhesive gave the highest bond strength in their tests.

Depending on the fabrication method, alternative types of adhesives could be used. Adhesive films can be used on FRP prepregs; however, the resin of the adhesive must be compatible with the resin of the FRP (Photiou et al. 2006). Because curing of the resins for both materials can occur simultaneously, this method not only hastens the strengthening process, but may also enhance molecular interlocking.

2.4 CAST/WROUGHT IRON, STEEL, AND ALUMINIUM

2.4.1 Cast/wrought iron

Cast iron was the earliest metallic material to be used structurally (from around 1780), and many examples of its use are still in service today (Hollaway and Teng 2008). Cast iron tends to be brittle and weak in tension with good wear resistance. It has been used as pipelines, beams in bridge decks, building columns, and bridge piers. There are several types of cast iron, namely, gray cast iron, malleable cast iron, ductile cast iron, and austenitic cast iron.

Gray cast iron is characterised by its graphitic microstructure, which causes fractures of the material to have a grey appearance. Malleable iron starts as a white iron casting that is then heat treated at about 900°C. A more recent development in the 1940s was nodular or ductile cast iron. Tiny amounts of magnesium or cerium added to these alloys slow down the growth of graphite precipitates by bonding to the edges of the graphite planes. Austenitic cast iron has a stable austenitic basic structure at ambient temperature.

Wrought iron superseded cast iron between the mid-1840s and early 1850s, as a high-performance structural material. It was eventually superseded by mild steel at the end of the 19th century. Wrought iron was produced from cast iron by raising it to a high temperature and subjecting it to a strong blast of air, which removed carbon and other impurities (Hollaway and Cadei 2002).

Material properties of some types of cast iron and wrought iron are listed in Table 2.3.

2.4.2 Steel

Structural steel sections are often made of steel plates or manufactured using the hot-rolled or cold-formed process. Typical sections include I-sections, channel sections, circular hollow sections (CHSs), rectangular hollow sections (RHSs), and LiteSteel beams (LSBs). Typical stress–strain curves

Table 2.3 Minimum values of yield stress and tensile strength for some types of cast iron and wrought iron

Name	Modulus of elasticity (GPa)	f_y (MPa)	f_u (MPa)
Gray cast iron	76–138	N/A	172
Malleable cast iron	172	228	359
Ductile cast iron	159	365	463
Wrought iron	193	159–221	234–372

Source: Lyons, W.C., and Plisga, G.J., *Standard Handbook of Petroleum and Natural Gas Engineering,* Oxford, Elsevier, 2006; Oberg, E. et al., *Machinery's Handbook,* 26th ed., New York, Industrial Press, 2000; Engineers Edge, Modulus of Elasticity, Strength Properties of Metals, 2011, http://www.engineersedge.com/manufacturing_spec/properties_of_metals_strength.htm (accessed January 16, 2011).

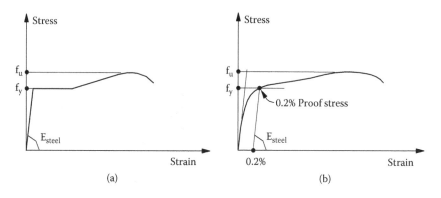

Figure 2.3 Stress–strain curves of metals (schematic view). (a) Mild steel or hot-rolled sections. (b) Cold-formed steel sections.

are plotted in Figure 2.3. For fabricated sections or hot-rolled sections an obvious yield plateau exists, whereas for cold-formed sections 0.2% proof stress is adopted to define the yield stress (Zhao et al. 2005).

Minimum values of yield stress and tensile strength are given in Tables 2.4 to 2.6, respectively, for typical steel plates, hot-rolled tubes, and cold-formed tubes used in Australia and Europe. The yield stress and tensile strength of LSB flange are 450 and 500 MPa, whereas the values are 380 and 490 MPa for LSB webs (LiteSteel Technology 2007). The modulus of elasticity of steel is about 200 GPa.

2.4.3 Aluminium

In Australia aluminium structural members are manufactured in accordance with AS/NZS 1734 (Standards Australia 1997a), AS/NZS 1866 (Standards Australia 1997b), and AS/NZS 1867 (Standards Australia 1997c). Minimum mechanical properties are listed in Table 3.3(A) of AS/

Table 2.4 Minimum values of yield stress and tensile strength for steel plates

(a) AS/NZS 3678 (Standards Australia 1996)

Grade	f_y (MPa)				
t (mm)	≤8	>8 but ≤12	>12 but ≤20	>20 but ≤32	f_u (MPa)
200	200	200	N/A	N/A	300
250	280	260	250	250	410
250L15	280	260	250	250	410
300 300L15	320	310	300	280	430
350 350L15	360	360	350	340	450
400 400L15	400	400	380	360	480
450 450L15	450	450	450	420	520 500
WR350 WR350L0	340	340	340	340	450

Source: Standards Australia, Structural Steel—Hot-Rolled Plates, Floor Plates, and Slabs, Australian/New Zealand Standard AS/NZS3678, Sydney, Australia, 1996.

(b) EN 10025 (2004)

Grade (for t ≤ 40 mm)	f_y (MPa)	f_u (MPa)
S235	235	360
S275	275	430
S355	355	510
S450	440	550

Source: EN 10025, Hot-Rolled Products of Structural Steels—Part 1: General Technical Delivery Conditions, EN 10025-1, European Committee for Standardization, Brussels, 2004.

Table 2.5 Minimum values of yield stress and tensile strength for hot-rolled tubes (EN 10210 2006)

Grade (for t ≤ 40 mm)	f_y (MPa)	f_u (MPa)
S235H	235	360
S275H	275	430
S355H	355	510
S275NH/NLH	275	390
S355NH/NLH	355	490
S420NH/NLH	420	540
S460NH/NLH	460	560

Source: EN 10210, Hot Finished Structural Hollow Sections of Non-Alloy and Fine Grain Steels—Part 1: Technical Delivery Requirements, EN 10210-1, European Committee for Standardization, Brussels, 2006.

Table 2.6 Minimum values of yield stress and tensile strength for cold-formed tubes

(a) AS 1163 (Standards Australia 1991)

Grade	f_y (MPa)	f_u (MPa)
C250, C250L0	250	320
C350, C350L0	350	430
C450, C450L0	450	500

Source: Standards Australia, Structural Steel Hollow Sections, Australian Standard AS 1163, Sydney, Australia, 1991.

(b) EN 10219 (2006)

Grade	f_y (MPa)		f_u (MPa)
CHS	t ≤ 16 mm	16 mm < t ≤ 40 mm	t ≤ 40 mm
RHS/SHS	t ≤ 16 mm	16 mm < t ≤ 24 mm	t ≤ 24 mm
S275NH S275NLH	275	265	370–510
S355NH S355NLH	355	345	470–630
S460NH S460NLH	460	440	550–720

Source: EN 10219, Cold Formed Welded Structural Hollow Sections of Non-Alloy and Fine Grain Steels—Part 1: Technical Delivery Requirements, EN 10219-1, European Committee for Standardization, Brussels, 2006.

NZS 1664.1 (Standards Australia 1997d). In Europe aluminium structural members are manufactured in accordance with EN 485-1 (1993) and EN 755-1 (1997). Minimum mechanical properties are listed in Table 3.2 of Eurocode 9 (EN 1999-1-1 2007). Some typical values are shown in Table 2.7 of this chapter. The stress–strain relationship of aluminium sections is similar to that of cold-formed sections in terms of the way yield stress is defined; i.e., 0.2% proof stress is used to define the yield stress (see Figure 2.3) because no obvious yield plateau exists.

2.5 FUTURE WORK

There is a lack of information on the shear stress and shear strain of FRP and adhesives. There is a need to study the effect of environmental conditions, such as temperatures, humidity, and UV light, on the material properties of FRP and adhesives. Some preliminary results (e.g., Al-Shawaf et al. 2005, Nguyen et al. 2011) are presented in Section 3.5. It is necessary to understand the influence of strain rate on the material properties of FRP and adhesives when dealing with FRP-strengthened metallic structures subject to impact and blast loading. Some preliminary results (e.g., Al-Zubaidy et al. 2012) are presented in Section 3.7.

Table 2.7 Minimum values of yield stress and tensile strength for typical aluminium sheet, plate, and pipe

(a) AS 1664.1 (1997)

Alloy	Temper	Product	Thickness (mm)	f_y (MPa)	f_u (MPa)
3004	H34	Sheet	<6.3	165	214
5083	H111	Extrusions	≤12	165	276
6061	T6	Pipe	≤25	241	290
6063	T5	Extrusions	≤12	110	152

Source: Standards Australia, Aluminium Structures—Part 1: Limit State Design, Australian/ New Zealand Standard AS/NZS1664.1, Sydney, Australia, 1997d.

(b) EC9 (2007)

Alloy	Temper	Product	Thickness (mm)	f_y (MPa)	f_u (MPa)
3004	H14	Sheet, strip, and plate	<6	180	220
5083	H14	Sheet, strip, and plate	≤25	280	340
6061	T6	Sheet, strip, and plate	≤12.5	240	290
6061	T6	Extruded open profile and extruded tube	≤20	240	260
6063	T5	Extruded tube	3 < t ≤ 25	110	160

Source: Eurocode 9, Design of Aluminium Structures—Part 1-1: General Structural Rules, EN1999-1-1, European Committee for Standardization, Brussels, 2007.

REFERENCES

Al-Shawaf, A., Al-Mahaidi, R., and Zhao, X.L. 2005. Tensile properties of CFRP laminates at subzero temperatures. In *Proceedings of the Australian Structural Engineering Conference*, Newcastle, UK, September 11–14, pp. 1–9.

Al-Zubaidy, H., Al-Mahaidi, R., and Zhao, X.L. 2012. Experimental investigation of bond characteristics between CFRP fabrics and steel plate joints under impact tensile loads. *Composite Structures*, 94(2), 510–518.

Cadei, J.M.C., Stratford, T.J., Hollaway, L.C., and Duckett, W.H. 2004. C595— *Strengthening metallic structures using externally bonded fibre-reinforced composites*. London: CIRIA.

El Damatty, A.A., and Abushagur, M. 2003. Testing and modelling of shear and peel behaviour for bonded steel/FRP connections. *Thin-Walled Structures*, 41(11), 987–1003.

EN 10025. 2004. *Hot-rolled products of structural steels. Part 1. General technical delivery conditions*. Brussels: European Committee for Standardization.

EN 10210. 2006. *Hot finished structural hollow sections of non-alloy and fine grain steels. Part 1. Technical delivery requirements*. Brussels: European Committee for Standardization.

EN 10219. 2006. *Cold formed welded structural hollow sections of non-alloy and fine grain steels. Part 1. Technical delivery requirements.* Brussels: European Committee for Standardization.

EN 1999-1-1. 2007. *Eurocode 9. Design of aluminium structures. Part 1-1. General structural rules.* Brussels: European Committee for Standardization.

EN 485-1. 1993. *Aluminium and aluminium alloys—Sheet, strip and plate. Part 1. Technical conditions for inspection and delivery.* Brussels: European Committee for Standardization.

EN 755-1. 1997. *Aluminium and aluminium alloys—Extruded rod/bar, tube and profiles. Part 1. Technical conditions for inspection and delivery.* Brussels: European Committee for Standardization.

Engineers Edge. 2011. Modulus of elasticity, strength properties of metals. http://www.engineersedge.com/manufacturing_spec/properties_of_metals_strength.htm (accessed January 16, 2011).

Fam, A., Witt, S., and Rizkalla, S. 2006. Repair of damaged aluminum truss joints of highway overhead sign structures using FRP. *Construction and Building Materials*, 20(10), 948–956.

Fawzia, S. 2007. Bond characteristics between steel and carbon fibre reinforced polymer (CFRP) composites. PhD thesis, Department of Civil Engineering, Monash University, Melbourne, Australia.

Fawzia, S., Al-Mahaidi, R., Zhao, X.L., and Rizkalla, S. 2007. Strengthening of circular hollow section steel tubular sections using high modulus CFRP sheets. *Construction and Building Materials*, 21(4), 839–845.

Fawzia, S., Zhao, X.L., Al-Mahaidi, R., and Rizkalla, S. 2005. Bond characteristics between CFRP and steel plates in double strap joints. *Advances in Steel Construction—An International Journal*, 1(2), 17–28.

Fernando, D., Yu, T., Teng, J.G., and Zhao, X.L. 2009. CFRP strengthening of rectangular steel tubes subjected to end bearing loads: Effect of adhesive properties and finite element modelling. *Thin-Walled Structures*, 47(10), 1020–1028.

Hollaway, L.C. 2010. A review of the present and future utilisation of FRP composites in the civil infrastructure with reference to their important in-service properties. *Construction and Building Materials*, 24(12), 2419–2445.

Hollaway, L.C., and Cadei, J. 2002. Progress in the technique of upgrading metallic structures with advanced polymer composites. *Progress in Structural Engineering and Materials*, 4(2), 131–148.

Hollaway, L.C., and Spencer, H. 2000. Modern developments. In *Manual of bridge engineering*, ed. Ryall, M.J., Parks, G.A.R., and Harding, J.E. London: Thomas Telford.

Hollaway, L.C., and Teng, J.C. 2008. *Strengthening and rehabilitation of civil infrastructures using fiber-reinforces polymers (FRP) composites.* Cambridge: Woodhead Publishing.

Hollaway, L.C., Zhang, L., Photiou, N.K., Teng, J.G., and Zhang, S.S. 2006. Advances in adhesive joining of carbon fibre/polymer composites to steel members for repair and rehabilitation of bridge structures. *Advances in Structural Engineering—An International Journal*, 9(6), 791–803.

Lanier, B., Schnerch, D., and Rizkalla, S. 2009. Behavior of steel monopoles strengthened with high modulus CFRP materials. *Thin-Walled Structures*, 47(10), 1037–1047.

LiteSteel Technology. 2007. *Design capacity tables for LiteSteel beam*. Sunnybank, Australia: LiteSteel Technology.

Lyons, W.C., and Plisga, G.J. 2006. *Standard handbook of petroleum and natural gas engineering*. Oxford: Elsevier.

Nguyen, T.C., Bai, Y., Zhao, X.L., and Al-Mahaidi, R. 2011. Mechanical characterization of steel/CFRP double strap joints at elevated temperatures. *Composite Structures*, 93(6), 1604–1612.

Oberg, E., Jones, F.D., Horton, H.L., and Ryffell, H.H. 2000. *Machinery's handbook*. 26th ed. New York: Industrial Press.

Pantelides, C.P., Nadauld, J., and Cercone, L. 2003. Repair of cracked aluminum overhead sign structures with glass fiber reinforced polymer composites. *Journal of Composites for Construction*, 7(2), 118–126.

Photiou, N.K., Hollaway, L.C., and Chryssanthopoulos, M.C. 2006. Selection of CFRP systems for structural upgrading. *Journal of Materials in Civil Engineering*, 18(5), 641–649.

Schnerch, D., Dawood, M., Rizkalla, S. and Sumner, E. 2007. Proposed design guidelines for strengthening steel bridges with FRP materials. *Construction and Building Materials*, 21(5), 1001–1010.

Sena Cruz, J.M., Barros, J.A.O., Gettu, R., and Azevedo, Á.F.M. 2006. Bond behavior of near-surface mounted CFRP laminate strips under monotonic and cyclic loading. *Journal of Composites for Construction*, 10(4), 295–303.

Seracino, R., Jones, N.M., Ali, M.S.M., Page, M.W., and Oehlers, D.J. 2007. Bond strength of near-surface mounted FRP strip-to-concrete joints. *Journal of Composites for Construction*, 11(4), 401–409.

Shaat, A., Schnerch, D., Fam, A., and Rizkalla, S. 2004. Retrofit of steel structures using fiber-reinforced polymers (FRP): State-of-the-art. Presented at Transportation Research Board (TRB) Annual Meeting, Washington, DC, January 11–15.

Standards Australia. 1991. *Structural steel hollow sections*. Australian Standard AS 1163. Sydney: Standards Australia.

Standards Australia. 1996. *Structural steel—Hot-rolled plates, floor plates and slabs*. Australian/New Zealand Standard AS/NZS 3678. Sydney: Standards Australia.

Standards Australia. 1997a. *Aluminium and aluminium alloys—Flat sheet, coiled sheet and plate*. Australian/New Zealand Standard AS/NZS 1734. Sydney: Standards Australia.

Standards Australia. 1997b. *Aluminium and aluminium alloys—Extruded rod, bar, solid and hollow shapes*. Australian/New Zealand Standard AS/NZS 1866. Sydney: Standards Australia.

Standards Australia. 1997c. *Aluminium and aluminium alloys—Drawn tubes*. Australian/New Zealand Standard AS/NZS 1867. Sydney: Standards Australia.

Standards Australia. 1997d. *Aluminium structures. Part 1. Limit state design*. Australian/New Zealand Standard AS/NZS 1664.1. Sydney: Standards Australia.

Tavakkolizadeh, M., and Saadatmanesh, H. 2001. Galvanic corrosion of carbon and steel in aggressive environments. *Journal of Composites for Construction*, 5(3), 200–210.

Wu, C., Zhao, X.L., Al-Mahaidi, R., and Duan, W.H. 2010. Experimental study on bond behaviour between UHM CFRP laminate and steel. Presented at the 5th International Conference on Composites in Civil Engineering, Beijing, September 27–29.

Zhao, X.L., and Zhang, L. 2007. State of the art review on FRP strengthened steel structures. *Engineering Structures*, 29(8), 1808–1823.

Zhao, X.L., Wilkinson, T., and Hancock, G.J. 2005. *Cold-formed tubular members and connections*. Oxford: Elsevier.

Chapter 3

Behaviour of the bond between FRP and metal

3.1 GENERAL

It is very important to understand the bond behaviour between fibre-reinforced polymer (FRP) and the substrate metal in order to study the debonding behaviour of FRP-strengthened metallic members. The bond between the FRP and metal is very different from that between FRP and concrete, which has been reported in Wu et al. (2002), Yuan et al. (2004), and Yao et al. (2005). The major differences have been described in Section 1.3.

Bond behaviour between FRP and steel has been widely studied experimentally and theoretically (Xia and Teng 2005, Fawzia et al. 2006, 2010, Wu et al. 2012a, Yu et al. 2012, Teng et al. 2012). The bond may lose at elevated temperature or due to dynamic loading (Liu et al. 2010, Nguyen et al. 2011). This chapter will address the method of bond testing, failure modes, the bond–slip model, and the effect of temperature and dynamic loading on bonds.

3.2 TESTING METHODS

3.2.1 Methods of bond test

Different bond testing methods were adopted by various researchers for different purposes of studying. The two commonly used methods are shown in Figure 3.1. They can be categorised as the following:

Type 1: When loading is directly applied to FRP (see Figure 3.1(a)).
Type 2: When loading is directly applied to a steel element with a gap in between (see Figure 3.1(b)).

In the type 1 testing method, either compressive force (El Damatty and Abushagur 2003) or tensile force (Xia and Teng 2005, Hart–Smith 1973) can be applied on FRP plates. The setup shown in Figure 3.1(a)(i) is only valid for testing bonds when the carbon fibre-reinforced polymer (CFRP) material is in compression. The disadvantages of applying compressive force in FRP include the local FRP failure due to its relatively smaller compression strength and the fact that FRP is generally utilised in tension. The other two cases in Figure 3.1(a) are more useful to study the bond between CFRP plates and steel. The setup shown in Figure 3.1(a)(ii), adopted by Xia

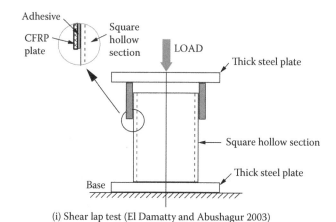

(i) Shear lap test (El Damatty and Abushagur 2003)

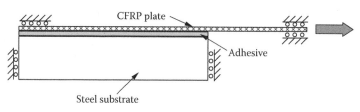

(ii) Single shear pull test (Xia and Teng 2005)

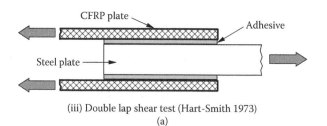

(iii) Double lap shear test (Hart-Smith 1973)

(a)

Figure 3.1 Testing methods to determine bond between steel and CFRP. (a) Type I: When loading is directly applied to FRP. (From Zhao, X. L., and Zhang, L., *Engineering Structures*, 29(8), 1808–1823, 2007.)

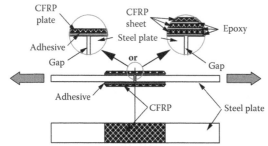

(i) Double shear pull test (Schnerch et al. 2004, Fawzia et al. 2006, Colombi and Poggi 2006)

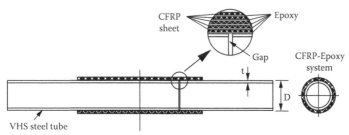

(ii) Single lap joint with circular hollow section (Jiao and Zhao 2004, Fawzia et al. 2007)

(b)

Figure 3.1 (Continued) Testing methods to determine bond between steel and CFRP. (b) Type 2: When loading is directly applied to steel element with a gap in between. (From Zhao, X.L., and Zhang, L., *Engineering Structures*, 29(8), 1808–1823, 2007.)

and Teng (2005), allows detailed monitoring and inspection of the failure process because only one path for debonding is possible. This method is also consistent with that used in studying the bond between FRP and concrete. However, it is a challenge to make sure that the alignment is maintained to minimise load eccentricity. This method may not be applicable for CFRP sheets due to the difficulty in gripping the sheets.

Double-shear pull tests in the type 2 testing method are often used (Schnerch et al. 2004, Fawzia et al. 2006, Colombi and Poggi 2006) to investigate the bond between steel plates and CFRP plates or sheets, as shown in Figure 3.1(b)(i). The single-lap joint in the type 2 testing method was used by Fawzia et al. (2007) and Jiao and Zhao (2004) to investigate the bond between CFRP sheets and steel tubes because it is not practical to apply CFRP on the inner surface of the steel tubes. This method is only applicable to CFRP sheets due to the curved steel surface. The major concern of this method is the uncertainty of the location of debonding failure. Four possible locations exist for the propagation of debonding. This makes the experimental instrumentation and observation more difficult. Attempts were made (e.g., in Fawzia et al. 2006) to use unequal bond lengths aiming to form debonding in the region with the shorter bond length. Mechanical clapping or transverse CFRP strengthening may also be applied to avoid debonding at one end.

Based on the above discussions, it is recommended that the test setup shown in Figure 3.1(a)(ii) be used for CFRP plates in establishing the bond–slip relationship between CFRP plates and steel in tension and that in Figure 3.1(b)(i) be used for CFRP sheets.

3.2.2 Methods of strain measurement

The strain distribution along the bond length can be measured using several strain gauges mounted on the surface of specimens. It can also be measured using the photogrammetry technique, e.g., noncontact 3D ARAMIS measurement system, which utilises the three-dimensional video correlation method and high-resolution digital CCD cameras (see Figure 3.2(a)). This system provides strain distribution over the whole area rather than at certain points, as the conventional strain gauging technique does. It has been proven that such noncontact measurement agrees well with those obtained from strain gauge readings (see Figure 3.2(b)).

3.3 FAILURE MODES

3.3.1 Typical failure modes

Possible failure modes in an FRP bonded steel system subjected to tensile force include the following:

1. Adhesion failure between steel and adhesive
2. Cohesion failure (adhesive layer failure)
3. Adhesion failure between FRP and adhesive
4. FRP delamination (separation of some carbon fibres from the resin matrix)
5. FRP rupture
6. Steel yielding

A schematic view of failure modes is given in Figure 3.3, while some examples are given in Figure 3.4. The failure mode in CFRP sheet specimens depends mainly on the tensile strength of the CFRP sheet (e.g., a normal-modulus CFRP sheet has higher tensile strength, whereas a high-modulus CFRP sheet has lower tensile strength, as shown in Fawzia et al. 2006, Jiao and Zhao 2004). The failure mode in CFRP plate specimens depends on the adhesive thickness and material properties (Xia and Teng

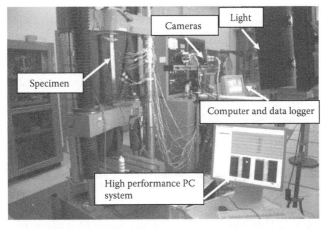

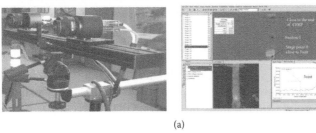

(a)

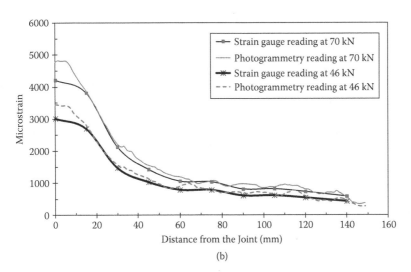

(b)

Figure 3.2 Strain measurement methods. (a) ARAMIS setup and screen capture. (b) Comparison of strain gauge readings and ARAMIS readings. (Courtesy of S. Fawzia, Monash University, Australia.)

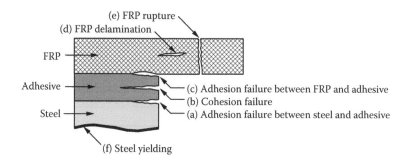

Figure 3.3 Schematic view of typical failure modes. (From Zhao, X.L., and Zhang, L., *Engineering Structures*, 29(8), 1808–1823, 2007.)

2005, Yu et al. 2012). Steel yielding (failure mode 6) is often avoided in the bond testing by using a sufficient plate thickness.

3.3.2 Key parameters affecting failure modes

As mentioned in Section 2.2.1, the terms *normal-modulus* and *high-modulus* FRP laminates are adopted in this book. The normal-modulus CFRP has an elastic modulus up to 250 GPa, whereas the high-modulus CFRP has an elastic modulus above 250 GPa.

For normal-modulus (e.g., 240 GPa) CFRP sheets the failure mode is a combination of mode 1 and mode 4. For high-modulus (e.g., 640 GPa) CFRP sheets, the failure mode is FRP rupture (mode 5). It should be noted that for multilayer CFRP sheets it is not practical to single out failure modes 2 and 3.

Xia and Teng (2005) studied the bond behaviour between normal-modulus CFRP plates and steel. Three different adhesives were adopted in the experimental program, as shown in Table 3.1. To achieve a wide range of values for the adhesive stiffness, the thickness of the adhesive layer was varied, as the elastic modulus of a commercially available adhesive cannot be readily modified. Four thicknesses were used: 1, 2, 4, and 6 mm. It should be noted that the first two thicknesses are realistic, but the last two thicknesses were used to achieve a wide range of the adhesive layer thickness. It was found that cohesion failure (mode 2) tends to occur for thin adhesives, whereas the failure mode changes to delamination of CFRP plates (mode 4) for thick adhesives, as shown in Figure 3.5. Mode 4 was found to be more brittle than mode 2. Failure modes 1 and 3 were not observed, indicating the strong bond capacity of the adhesives to the roughened steel and the FRP plate surfaces. For high-modulus CFRP plates (Wu et al. 2012a) the failure modes tend to be 2 and 4 for short bond length (see Figure 3.4(a) and (b)), whereas the FRP rupture (mode 5) occurs when the bond length is long enough.

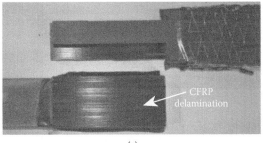

(a)

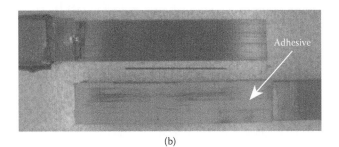

(b)

(c)

Figure 3.4 Examples of failure modes in double-shear pull tests. (a) CFRP delamination (mode 4). (Courtesy of C. Wu, Monash University, Australia.) (b) Cohesion failure (mode 2). (Courtesy of C. Wu, Monash University, Australia.) (c) CFRP rupture (mode 5). (Courtesy of H.B. Liu, Monash University, Australia.)

Table 3.1 Material properties of adhesives

Adhesive	Tensile strength $f_{t,a}$ (MPa)	Elastic modulus E_a (MPa)	Poisson's ratio ν_a	Ultimate tensile strain (%)
A	22.53	4013	0.36	0.561
B	20.48	10793	0.27	0.190
C	13.89	5426	0.31	0.256

Source: Xia, S.H., and Teng, J.G., Behavior of FRP-to-Steel Bond Joints, in *Proceedings of International Symposium on Bond Behaviour of FRP in Structures (BBFS 2005)*, Hong Kong, December, 2005, pp. 419–426.

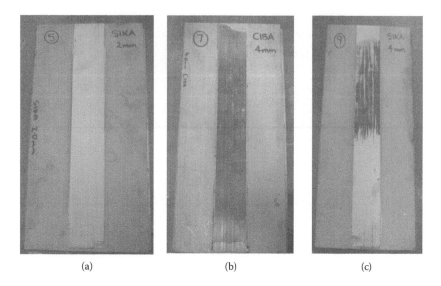

(a)   (b)   (c)

Figure 3.5 Example of failure modes in single-shear pull tests. (a) Cohesion failure (mode 2). (b) CFRP delamination (mode 4). (c) Combination of cohesion failure and CFRP delamination (modes 2 and 4). (Courtesy of J.G. Teng, Hong Kong Polytechnic University, China.)

3.4 BOND–SLIP MODEL

3.4.1 Strain distribution

Strain distribution along the bond length provides important information to understand the bond between CFRP and steel. As discussed in Section 3.2.2, strain distributions could be captured by using strain gauges or the noncontact photogrammetry technique. Some typical strain distributions are presented in Figure 3.6(a) from single-shear pull tests (see Figure 3.1(a) (ii)) and in Figure 3.6(b) from double-shear pull tests (see Figure 3.1(b)(i)), where P_{ult} is the ultimate capacity (i.e., bond strength). It is clear that the strain increases along the bond length as the applied load increases.

When multiple layers of CFRP are applied to the surface of steel, the strain distribution is not uniform across the layers. Some typical strain distributions across three and five layers of CFRP sheets are shown in Figure 3.7. The strain gauges were located at the joint of the double-shear pull test specimens. The strain values reduce almost linearly from layer 1 (which is closest to the steel surface) to layer 3. There is little reduction from layer 3 to layer 5. It seems that the benefit of applying more than three layers of CFRP in strengthening becomes less significant because of the reduced contributions from layer 4 and beyond.

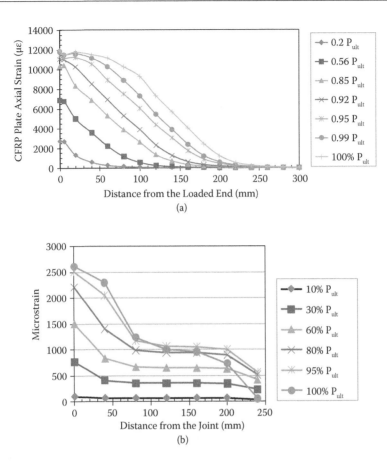

Figure 3.6 Typical strain distribution along bond length (cohesion failure). (a) Single-shear pull test when $P < P_{ult}$. (Adapted from Yu, T. et al., *Composites Part B: Engineering*, 43(5), 2279–2289, 2012.) (b) Double-shear pull test. (From Wu, C. et al., *Thin-Walled Structures*, 51(2), 147–157, 2012.)

3.4.2 Bond–slip curves

The interfacial shear behaviour of a bonded joint is often characterised by the so-called bond–slip curve, which depicts the relationship between the local interfacial shear stress and the relative slip between the two adherends. In a single-shear pull test, the interfacial shear stress and the slip can be found from strain distributions described in Section 3.4.1 using Equation (3.1) as derived in Pham and Al-Mahaidi (2005).

$$\tau_{i+\frac{1}{2}} = \frac{(\varepsilon_i - \varepsilon_{i-1})}{(L_i - L_{i-1})} E_{CFRP} t_{CFRP\_plate} \tag{3.1a}$$

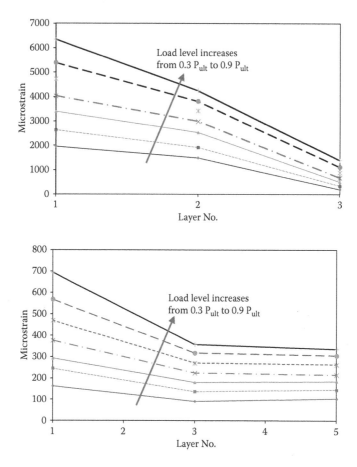

Figure 3.7 Distribution of strain across the layers. (a) Normal-modulus CFRP sheet. (Adapted from Liu, H.B. et al., *International Journal of Structural Stability and Dynamics*, 10(1), 1–20, 2010.) (b) High-modulus CFRP sheet. (Adapted from Fawzia, S., Bond Characteristics between Steel and Carbon Fibre Reinforced Polymer (CFRP) Composites, PhD thesis, Department of Civil Engineering, Monash University, Melbourne, Australia, 2007.)

$$\delta_{i+\frac{1}{2}} = \frac{(L_i - L_{i-1})(\varepsilon_i + \varepsilon_{i-1})}{4} + \frac{(L_{i-1} - L_{i-2})(\varepsilon_{i-1} + \varepsilon_{i-2})}{2}$$

$$+ \sum_{i=3}^{i} \frac{(L_{i-2} - L_{i-3})(\varepsilon_{i-2} + \varepsilon_{i-3})}{2}$$

(3.1b)

where ε_i is the reading of the ith strain gauge counted from the free end of the CFRP plate, with $\varepsilon_0 = 0$; L_i is the distance of the ith strain gauge from the free end of the CFRP plate, with $L_0 = 0$; E_{CFRP} and t_{CFRP_plate} are the elastic modulus and thickness of the CFRP plate, respectively; and $\tau_{i+1/2}$ and $\delta_{i+1/2}$ are the shear stress and slip at the middle point between the ith strain gauge and the i – 1th strain gauge.

Similarly for a double-shear pull test setup, the interfacial shear stress and the slip can be estimated from strain gauge readings using Equation (3.2) (Wu et al. 2012a).

$$\tau_{i+\frac{1}{2}} = \frac{(\varepsilon_{i+1} - \varepsilon_i)}{(L_{i+1} - L_i)} E_{CFRP} t_{CFRP\_plate} \qquad (3.2a)$$

$$\delta_{i+\frac{1}{2}} = \frac{1}{2}(L_{i+1} - L_i)(\varepsilon_{i+1} + \varepsilon_i) \qquad (3.2b)$$

As shown in Figure 2.2, adhesives could have different strain-stress behaviours; i.e., some are almost linear (called linear adhesive in this book), and some demonstrate significant nonlinearity (called nonlinear adhesive in this book). The influence of different types of adhesives on bond behaviour was studied by Yu et al. (2012) by adopting one linear adhesive and one nonlinear adhesive, where a single-shear pull test setup (see Figure 3.1(a) (ii)) was used.

Figure 3.8 shows the interfacial shear stress distributions along the CFRP plate at different stages of loading/deformation for a specimen with a linear adhesive (shown in Figure 3.8(a) and (b)) and a specimen with a nonlinear

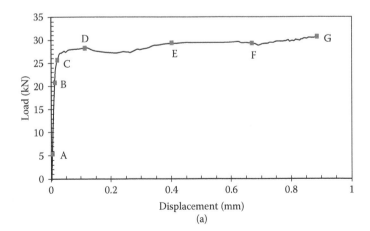

Figure 3.8 Typical shear stress distribution along bond length. (a) Load-displacement curve for a linear adhesive. (From Yu, T. et al., *Composites Part B: Engineering*, 43(5), 2279–2289, 2012).

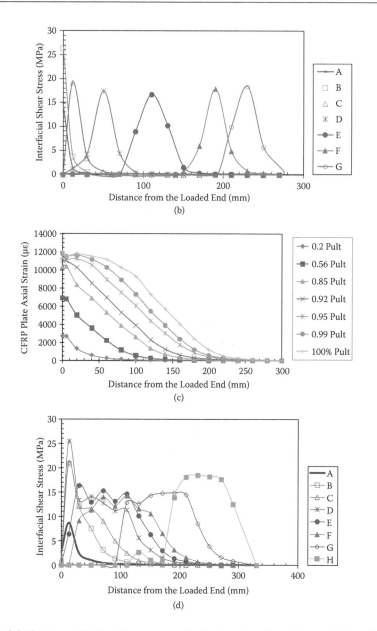

Figure 3.8 (Continued) Typical shear stress distribution along bond length. (b) Interfacial shear stress distribution for a linear adhesive. (c) Load-displacement curve for a nonlinear adhesive. (d) Interfacial shear stress distribution for a nonlinear adhesive. (From Yu, T. et al., *Composites Part B: Engineering,* 43(5), 2279–2289, 2012).

adhesive (shown in Figure 3.8(c) and (d)), respectively. The interfacial shear stresses were obtained from the axial strains of the CFRP plate using Equation (3.1).

Figure 3.8(a) and (b) shows that the development of shear stresses in the specimen with a linear adhesive includes three distinctive stages: (1) initial stage (representative points A and B), when the shear stress is the largest at or very close to the loaded plate end and reduces gradually toward the free end of the plate; (2) softening stage (representative point C), when the shear stress near the loaded end starts to decrease after it has reached its maximum value (i.e., the loaded end is on the descending branch of the bond–slip curve); the load resisted by the bonded joint continues to increase in this stage, but the load-displacement curve becomes nonlinear; and (3) debonding stage (representative points D to G), after the shear stress at the loaded end has reduced to zero; the peak shear stress moves gradually toward the free end and the load is almost constant during this stage.

Figure 3.8(d) shows the interfacial shear stress distributions in the specimen with a nonlinear adhesive. These shear stress distributions are quite different from those shown in Figure 3.8(b) for the case with a linear adhesive. In this specimen, localised debonding occurred near the loaded end at a low load due to a local stress concentration there. The localised debonding explains why the maximum of the interfacial shear stress does not appear at the loaded end (Figure 3.8(d)). A similar phenomenon was noted by Yuan et al. (2004) for CFRP-to-concrete bonded joints. At many loading/deformation levels, a certain length of the bonded interface is subject to similar interfacial shear stresses, which are approximately equal to the peak shear bond stress. In particular, after the shear stress at the loaded end has reached its peak value, this stress value is maintained and a stress plateau develops as the load increases until a certain load level at which the loaded end enters the softening state and the constant stress region has been fully developed. When the interfacial stress at the loaded end reaches zero, the stress plateau starts to move gradually from the loaded end to the free end until point H of the load-displacement curve (see Figure 3.8(c)) is reached.

For both linear and nonlinear adhesives, Figure 3.8 shows that significant shear stresses were developed only within a limited region of the bonded interface at any instance. After the shear stress at the loaded end has decreased to zero, this high-stress region starts to move gradually toward the free end, but the size of the high-stress region remains almost constant. As the tensile force that the CFRP plate can take relies on the shear stress transfer across the bonded interface, this observation clearly explains the existence of an effective bond length, beyond which any further increase in the bond length does not lead to a further increase in the bond strength, but leads to an increase in the ductility of the failure process.

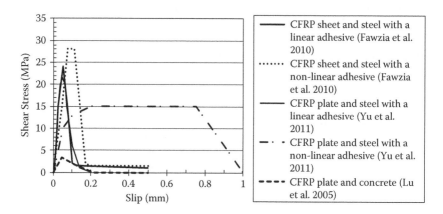

Figure 3.9 Examples of bond–slip curves.

Some examples of bond–slip curves are given in Figure 3.9 for CFRP–steel systems and for a CFRP-concrete system. It is obvious that the fracture energy (area under the curve) for CFRP–steel systems is much higher than that for the CFRP-concrete system, as expected. The difference between the curves with linear adhesives and those with nonlinear adhesives is distinct for the CFRP plate–steel system, whereas the difference is not significant for the CFRP sheet-steel system. There is a similarity in bond–slip curves for the CFRP sheet and CFRP plate with linear adhesives.

3.4.3 Bond–slip model

A bilinear bond–slip model was firstly proposed by Xia and Teng (2005). For simplicity purposes, a single bilinear bond–slip model was adopted by Fawzia et al. (2010) for a CFRP sheet–steel system with linear and nonlinear adhesives. A schematic view of such a model is shown in Figure 3.10(a), which can be determined by three key parameters: τ_f (maximum shear stress), δ_1 (initial slip), and δ_f (maximum slip). Calibration of these key parameters was carried out by Fawzia et al. (2010) using finite element simulation and parametric studies. They are summarised as follows:

$$\tau_f = f_{t,a} \tag{3.3a}$$

$$\delta_1 = \frac{t_a}{10} \tag{3.3b}$$

$$\delta_f = \frac{t_a}{4} \text{ for } t_a = 0.1 \text{ to } 0.5 \text{mm} \tag{3.3c}$$

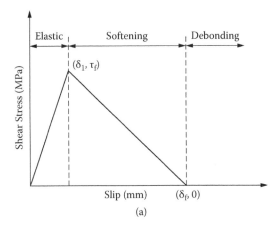

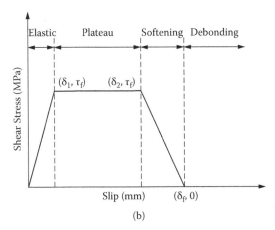

Figure 3.10 Schematic view of bond–slip models. (a) Bond–slip model for CFRP sheet and steel (with either linear or non-linear adhesives) and for CFRP plate and steel (with linear adhesives). (b) Bond–slip model for CFRP plate and steel (with non-linear adhesives). (Adapted from Xia, S. H., and Teng, J. G., 2005; Fawzia, S. et al. 2010; Yu, T. et al. 2012.)

$$\delta_f = 0.125 + \frac{t_a - 0.5}{10} \quad \text{for } t_a = 0.5 \text{ to } 1\,\text{mm} \tag{3.3d}$$

$$t_a = \frac{\frac{1}{2} \cdot (T - t_{steel})}{n} - t_{CFRP\_sheet} \tag{3.3e}$$

where $f_{t,a}$ is the tensile strength of adhesive, as described in Section 2.3; t_a is the thickness of adhesive, which is assumed to be the adhesive between the steel plate and the first layer of CFRP; t_{steel} is the thickness of the steel plate; t_{CFRP_sheet} is the thickness of one layer of CFRP sheet; T is the average total thickness of the specimen at the joint; and n is the number of CFRP layers on one side of the joint.

For the CFRP plate–steel system with linear adhesives, the shape of the bond–slip curve is very much the same as that for the CFRP sheet–steel system. Hence, a bilinear bond–slip model can be adopted. It is logical to adopt a similar expression of the three key parameters, as in Equation (3.3). Based on the limited test results in Yu et al. (2012), the three key parameters could be approximately estimated as follows:

$$\tau_f = 0.9 f_{t,a} \tag{3.4a}$$

$$\delta_1 = \frac{t_a}{40} \tag{3.4b}$$

$$\delta_f = \frac{2t_a}{15} \tag{3.4c}$$

where t_a is the adhesive thickness between CFRP plate and steel.

For the CFRP plate–steel system with nonlinear adhesives, the bond–slip curves look like a trapezoid with an obvious plateau. A simplified model is proposed in Figure 3.10(b), which can be determined by four key parameters: τ_f (maximum shear stress), δ_1 (initial slip), δ_2 (slip at the end of plateau), and δ_f (maximum slip). The parameters can be approximated in a similar format as in Eqs. (3.3) and (3.4) based on the limited test results in Yu et al. (2012). They are presented in Equation (3.5).

$$\tau_f = f_{t,a} \tag{3.5a}$$

$$\delta_1 = \frac{t_a}{10} \tag{3.5b}$$

$$\delta_2 = \frac{3t_a}{4} \tag{3.5c}$$

$$\tau_f = t_a \tag{3.5d}$$

3.4.4 Estimation of bond strength and effective bond length

3.4.4.1 Hart-Smith (1973) model and Xia and Teng (2005) model for bond between CFRP plate and steel

In this section, two theoretical models are presented for bond between CFRP plate and steel, namely, the Hart-Smith model (Hart-Smith 1973) and Xia and Teng model (Xia and Teng 2005).

Hart-Smith (1973) analysed adhesively bonded double-shear pull test joints to estimate the ultimate load (bond strength). In the Hart-Smith model, the maximum bond strength of such a joint, with bond length longer than effective bond length, is given by

$$P_{ult} = b_{CFRP} \cdot \min\{P_i, P_o\} \qquad (3.6)$$

where b_{CFRP} is the width of CFRP, and P_i and P_0 are the bond strengths per unit width of the joint:

$$P_i = \sqrt{2\tau_f t_a \left(\frac{1}{2}\gamma_e + \gamma_p\right)2E_{steel}t_{steel}\left(1+\frac{E_{steel}t_{steel}}{2E_{CFRP}t_{CFRP\_plate}}\right)}$$

(if $E_{steel}t_{steel} < 2E_{CFRP}t_{CFRP\_plate}$) $\qquad (3.7a)$

$$P_o = \sqrt{2\tau_f t_a \left(\frac{1}{2}\gamma_e + \gamma_p\right)4E_{CFRP}t_{CFRP\_plate}\left(1+\frac{2E_{CFRP}t_{CFRP\_plate}}{E_{steel}t_{steel}}\right)}$$

(if $E_{steel}t_{steel} \geq 2E_{CFRP}t_{CFRP\_plate}$) $\qquad (3.7b)$

The effective bond length is given by

$$L_e = \frac{P_{ult}}{2\tau_f b_{CFRP}} + \frac{2}{\lambda} \qquad (3.8)$$

where

$$\lambda = \sqrt{\frac{G_a}{t_a}\left(\frac{1}{E_{CFRP}t_{CFRP\_plate}} + \frac{2}{E_{steel}t_{steel}}\right)} \qquad (3.9)$$

where E and t are the modulus and thickness, τ_f is the adhesive shear strength, γ_e (defined as τ_f/G_a) and γ_p are the adhesive elastic and plastic shear strains, respectively, G_a is the shear modulus of the adhesive, γ_p may be taken as $3\gamma_e$ for normal-modulus CFRP plate (Liu et al. 2005), and γ_p may be taken as $5\gamma_e$ for high-modulus CFRP plate (Wu et al. 2012a).

For the specimens with bond length (L) shorter than L_e, the bond strength can be estimated by

$$P_{ult,L<L_e} = P_{ult} \cdot \frac{L}{L_e} \qquad (3.10)$$

Wu et al. (2012a) proved that the above equations are also valid for bonds between high-modulus CFRP plates and steel.

In the Xia and Teng model (Xia and Teng 2005, Yu et al. 2012), the bond strength of the specimen, with bond length longer than effective bond length, is determined by

$$P_{ult} = b_{CFRP}\sqrt{2E_{CFRP}t_{CFRP\_plate}G_f} \qquad (3.11)$$

where G_f is the interfacial fracture energy defined by the area under the bond–slip curve in Figure 3.10, i.e.,

$$G_f = \frac{1}{2}\tau_f\delta_f \text{ for linear adhesives} \qquad (3.12a)$$

$$G_f = \frac{1}{2}\tau_f(\delta_f + \delta_2 - \delta_1) \text{ for nonlinear adhesives} \qquad (3.12b)$$

In the bond–slip model for the CFRP plate–steel system with linear adhesives (see Figure 3.10(a)), δ_1 is generally very small compared to values of δ_f. The effective bond length can be approximated by the analytical expression for a bond–slip model with a rigid ascending branch followed by a linearly descending branch, which is given by (Yuan et al. 2004)

$$L_e = \frac{\pi}{2\sqrt{\tau_f/E_{CFRP}t_{CFRP\_plate}\delta_f}} \qquad (3.13)$$

3.4.4.2 Modified Hart-Smith model (Fawzia et al. 2006) for bond between CFRP sheets and steel

The Hart-Smith model presented in Section 3.4.4.1 was derived for the double-shear pull test joints with one layer of outside adherend. When more than one layer of CFRP sheets are applied, some modifications are necessary before using the Hart-Smith model. A modified Hart-Smith model (MHSM) was developed in Fawzia et al. (2006). It was assumed that the adhesive to be used for the model is the adhesive between the steel plate and the first layer of CFRP. The rest of the material above this layer of adhesive is considered the outside adherend. It has the same expression as those in Eqs. (3.6) to (3.10), except for t_a and t_{CFRP_plate}. The thicknesses t_a and t_{CFRP_plate} can be determined as follows, based on equal-thickness epoxy between the CFRP and steel and between each of the CFRP layers:

$$t_a = t_{a\_layer} = \frac{\frac{1}{2} \cdot (T - t_{steel})}{n} - t_{CFRP\_sheet} \tag{3.14a}$$

$$t_{CFRP\_plate} = n \cdot t_{CFRP\_sheet} + (n - 1) \cdot t_{a\_layer} \tag{3.14b}$$

where all the parameters are the same as those defined in Equation 3.3(e).

3.5 EFFECT OF TEMPERATURE ON BOND STRENGTH

3.5.1 Influence of subzero temperature on bond strength

Al-Shawaf et al. (2005, 2006) and Al-Shawaf and Zhao (2013) studied the effect of subzero temperatures (0, −20, and −40°C) on material properties of CFRP and adhesives, and the bond behaviour between CFRP and steel. A normal-modulus CFRP sheet and three types of adhesives (Araldite 420, MBrace saturant, Sikadur 30) were adopted in the testing program.

An increased brittleness of the CFRP sheet was observed as the temperature decreased, as illustrated in Figure 3.11. The main feature in the failure mechanism of the subzero-tested specimens is the brittle mode characterised by a series of continuous material fragmentation and splitting associated with sudden breakage of the fibres. The tensile elastic modulus,

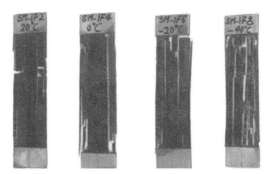

Figure 3.11 Effect of subzero temperature on failure mode of CFRP sheet. (Courtesy of A. Al-Shawaf, Monash University, Australia.)

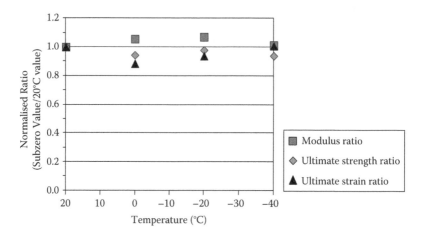

Figure 3.12 Effect of subzero temperature on material properties of CFRP sheet. (Adapted from Al-Shawaf, A. et al., Tensile Properties of CFRP Laminates at Subzero Temperatures, in *Proceedings of the Australian Structural Engineering Conference*, Newcastle, UK, September 11–14, 2005, pp. 1–9.)

ultimate strength, and strain of the CFRP sheet do not significantly vary (within 10%, as shown in Figure 3.12) as the environmental temperature decreases from 20°C to −40°C.

There is a general increase in modulus of elasticity of three adhesives tested when temperature decreases from 20°C to −40°C (Al-Shawaf and Zhao 2013). The increased percentage is about 50% for Araldite 420, 90% for MBrace saturant, and 15% for Sikadur 30. An increase of about 45% and a reduction of about 40% in the ultimate strength was found for Sikadur 30 and MBrace saturant, respectively, whereas there was nearly no change for Araldite 420. The ultimate strain decreases as the temperature decreases for Araldite 420 (about 60% reduction) and MBrace saturant (about 80%). The influence of subzero temperature on the ultimate strain is minimal for Sikadur 30.

It was found by Al-Shawaf and Zhao (2013) that no reduction in bond strength was found at subzero temperatures down to −40°C for specimens with Araldite 420 and Sikadur 30. However, the bond strength reduces about 40% for specimens with MBrace saturant when the temperature drops from 20°C to −40°C.

3.5.2 Influence of elevated temperature on bond strength

It was shown in Section 3.5.1 that subzero temperatures do not influence the bond between CFRP and steel. However, the bond strength may reduce significantly when the temperature approaches the glass transition temperature (usually called T_g), beyond which the stiffness of adhesives reduces (Hollaway and Teng 2008).

The glass transition temperature can be determined using several experimental techniques, such as differential scanning calorimetry (DSC), thermomechanical analysis (TMA), and dynamic mechanical thermal analysis (DMTA). The elastic modulus, heat flow, or volume change is measured against elevated temperatures. A schematic view of estimating T_g is presented in Figure 3.13. It should be pointed out that the value of T_g depends on the definition adopted since the transition temperature from a glassy state to a rubbery state lies within a range, e.g., between $T_{g,lower}$ and $T_{g,upper}$.

Nguyen et al. (2011) performed a series of double-shear pull tests on steel-CFRP joints with different bond lengths at temperatures across the glass transition of the adhesive. The T_g was found to be 42°C. It was determined by the intersection point of two tangent lines; one was drawn near the $T_{g,lower}$ point and the other was drawn near the centre between $T_{g,lower}$ and $T_{g,upper}$. The T_g defined in such a way is closer to $T_{g,lower}$ than to $T_{g,upper}$.

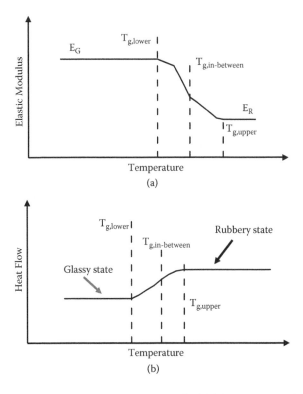

Figure 3.13 Schematic view of methods to measure T_g. (a) Schematic view of elastic modulus versus temperature. (b) Schematic view of heat flow versus temperature.

At a temperature lower than T_g, the joints failed through CFRP delamination (mode 4 shown in Figure 3.3). When the temperature increased to near or greater than T_g, the joints failed through cohesion failure (mode 2 shown in Figure 3.3). The change of failure modes was due to the difference in temperature dependence of the adhesive layer and CFRP adherend, as the former degrades much faster than the latter.

Figure 3.14 shows the influence of elevated temperature on bond strength. The experimental joint strength dropped by about 15, 50, and 80% when temperatures reached T_g, 10°C above T_g, and 20°C above T_g, respectively. The decrease of joint ultimate load with temperature was found to be

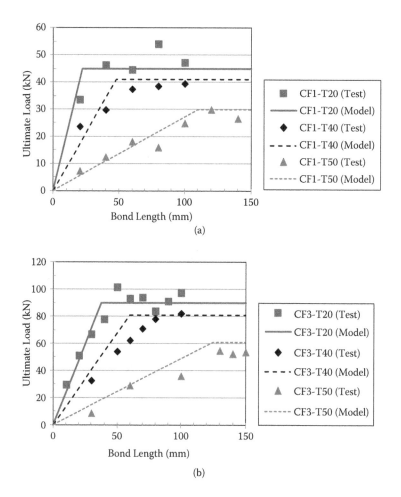

Figure 3.14 Influence of elevated temperature on bond strength. (a) One-layer CFRP sheet. (b) Three layers of CFRP sheets. (Adapted from Nguyen, T.C. et al., *Composite Structures*, 93(6), 1604–1612, 2011.)

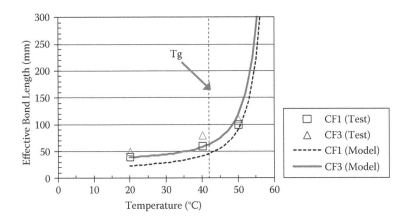

Figure 3.15 Influence of elevated temperature on effective bond length. (Adapted from Nguyen, T.C. et al., *Composite Structures*, 93(6), 1604–1612, 2011.)

related to the shear strength degradation of the adhesive as well as the change of the effective bond length.

Figure 3.15 shows the influence of elevated temperature on the effective bond length. The effective bond length of adhesively bonded steel-CFRP double-shear pull test joints was found experimentally to increase with temperature. The effective bond length at 40°C (near to T_g) was about twice that at room temperature. It was further found that the temperature-induced reduction of the joint ultimate load is more prominent when the bond length is less than the effective value, suggesting that a bond length larger than the effective value should be used in joint ultimate strength design to account for the fact that the joint may be exposed to elevated temperatures.

Figure 3.16 shows the influence of elevated temperature on joint stiffness normalised by that at 20°C. The joint stiffness significantly decreased as temperature increased: a 20% reduction at the T_g, 50% at 10°C above T_g, and 80% at 20°C above T_g. The results of DMTA tests on the same adhesive are also plotted in Figure 3.16. It can be clearly seen that a very similar temperature dependence trend is found. Because the joint displacement is mainly caused by the shear deformation of the adhesive layer (for failure mode 2), it seems that the stiffness reduction of the steel-CFRP composite double-shear pull test joints under elevated temperatures is caused by that of the adhesive layer.

3.5.3 Theoretical analysis of effect of elevated temperature on bond

Nguyen et al. (2011) developed a mechanism-based modelling by incorporating the temperature-dependent mechanical properties of the adhesive

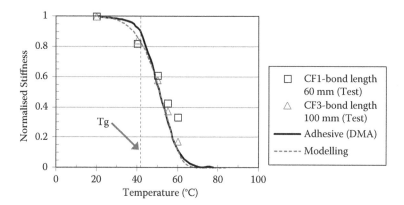

Figure 3.16 Influence of elevated temperature on stiffness. (Adapted from Nguyen, T.C. et al., *Composite Structures*, 93(6), 1604–1612, 2011.)

into the Hart-Smith model to describe the change of the effective bond length and the joint stiffness and strength.

The mechanical properties of the adhesive can be described as follows (Bai et al. 2008, Bai and Keller 2011):

$$C_m = C_{initial} (1 - \alpha_g) + C_{end} \alpha_g \tag{3.15}$$

where C_m is the resin-dominated material property such as bending modulus, shear modulus, or shear strength. $C_{initial}$ is the property at initial state (ambient temperature); C_{end} is the property at final state (after the glass transition), being zero because the DMTA test revealed that after glass transition the modulus of the epoxy was negligible. α_g is conversion degree of the glass transition, determined by the following Arrhenius equation:

$$\frac{d\alpha_g}{dT} = \frac{A_g}{\beta} \exp\left(\frac{-E_{A,g}}{RT}\right)(1 - \alpha_g) \tag{3.16}$$

where A_g is the preexponential factor, β is the constant heating rate (e.g., 2°C/min in Nguyen et al. 2011), $E_{A,g}$ is the activation energy (which is a constant for the glass transition), T is the temperature (K), R is the universal gas constant (8.314 J/mol.K), and A_g and $E_{A,g}$ are kinetic parameters with values of 1.34 × 10^18/min and 115.36 kJ/mol, respectively. The predicted temperature-dependent elastic modulus of Alradite 420 is plotted in Figure 3.16, which compares well with the DMTA results.

According to Hart-Smith (1973), at a certain temperature T, the ultimate load $P_{ult}(T)$ is given by

$$P_{ult}(T) = b_{CFRP} \cdot \min\{P_i(T), P_o(T)\} \qquad (3.17)$$

$$P_i(T) = \sqrt{2\tau_f(T)t_a\left(\frac{1}{2}\gamma_e(T) + \gamma_p(T)\right)2E_{steel}t_{steel}\left(1 + \frac{E_{steel}t_{steel}}{2E_{CFRP}t_{CFRP\_plate}}\right)}$$

$$(E_{steel}t_{steel} < 2E_{CFRP}t_{CFRP\_plate}) \qquad (3.18a)$$

$$P_o(T) = \sqrt{2\tau_f(T)t_a\left(\frac{1}{2}\gamma_e(T) + \gamma_p(T)\right)4E_{CFRP}t_{CFRP\_plate}\left(1 + \frac{2E_{CFRP}t_{CFRP\_plate}}{E_{steel}t_{steel}}\right)}$$

$$(E_{steel}t_{steel} \geq 2E_{CFRP}t_{CFRP\_plate}) \qquad (3.18b)$$

where $\tau_f(T)$ is the temperature-dependent adhesive shear strength. E_{steel}, t_{steel} and E_{CFRP}, t_{CFRP_plate} represent the elastic modulus and thickness for inside (steel) and outside (CFRP) adherend, respectively. The values of t_a and t_{CFRP_plate} can be estimated using Equation (3.14). The temperature-dependent elastic modulus of steel can be calculated from most of the steel structure standards (e.g., AS 4100 (Standards Australia 1998)). It can be proven that there is no obvious reduction (less than 1%) in modulus for steel (Standards Australia 1998) and CFRP (Sauder et al. 2004) at 60°C. Therefore, both E_{steel} and E_{CFRP} can be considered temperature independent. $\gamma_e(T)$ and $\gamma_p(T)$ are temperature-dependent elastic and plastic shear strains for adhesive that can be estimated using Eqs. (3.19) and (3.20), respectively:

$$\gamma_e(T) = \frac{\tau_f(T)}{G_a(T)} \qquad (3.19)$$

$$\gamma_p(T) = 3\gamma_e(T) \qquad (3.20)$$

where $G_a(T)$ is the temperature-dependent adhesive shear modulus.

Knowing $P_{ult}(T)$, the joint effective bond length, $L_e(T)$, is calculated as

$$L_e(T) = \frac{P_{ult}(T)}{2\tau_f(T)b_{CFRP}} + \frac{2}{\lambda(T)} \qquad (3.21a)$$

where

$$\lambda(T) = \sqrt{\frac{G_a(T)}{t_a}\left(\frac{1}{E_{CFRP}t_{CFRP\_plate}} + \frac{2}{E_{steel}t_{steel}}\right)} \qquad (3.21b)$$

The predicted temperature-dependent effective bond length is plotted in Figure 3.15 and agrees reasonably well with the experimental data.

For the specimens with bond length (L) shorter than $L_e(T)$, the bond strength can be estimated by

$$P_{ult,L<L_e}(T) = P_{ult}(T) \cdot \frac{L}{L_e(T)} \qquad (3.22)$$

The predicted bond strength versus bond length curves is plotted in Figure 3.14, and good agreement with test data is achieved.

3.6 EFFECT OF CYCLIC LOADING ON BOND STRENGTH

Matta (2003) and Liu et al. (2010) conducted tests to understand the influence of fatigue loading on the bond between steel and CFRP. CFRP plate was used by Matta (2003), whereas normal-modulus (240 GPa) and high-modulus (640 GPa) CFRP sheets were used by Liu et al. (2010). The specimens were tensioned to failure after enduring a preset number of fatigue cycles that ranged from 0.5 to 10 million at different load ratios ranging from 0.15 to 0.55. The load ratio is defined as the ratio of the maximum value of the applied load (P_{max}) to its static ultimate strength (F_1). Meanwhile, the specimens with the same configurations are tensioned to failure under static load to obtain reference values. By comparing these results, i.e., F_1 and F_2, where F_2 is the ultimate strength after a preset number of fatigue cycles, the influence of fatigue loading can be obtained.

The relationships between the load ratio (P_{max}/F_1), bond strength ratio (F_2/F_1), and preset number of fatigue cycles (N) are shown in Figure 3.17 for normal-modulus CFRP specimens. It can be seen that for normal-modulus CFRP, the reduction in bond strength is around 20 to 30% even when the load ratio is 0.3 with the preset fatigue cycle of 8 million. For joints with high-modulus CFRP sheets, it was found that the fatigue loading does not affect the bond strength even when the load ratio is as high as 0.55 and the number of fatigue cycles is up to 10 million. The fatigue cycles do not change the failure modes; i.e., they are very much the same as those observed in static bond tests without fatigue cycles.

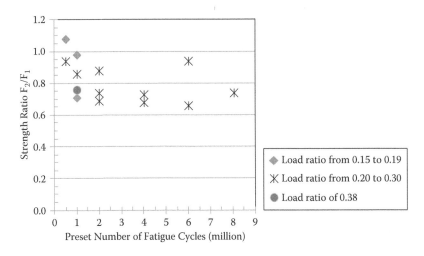

Figure 3.17 Influence of fatigue loading on bond strength. (Adapted from Liu, H.B. et al., *International Journal of Structural Stability and Dynamics*, 10(1), 1–20, 2010.)

Wu et al. (2012b) studied the influence of fatigue cycle on the bond between high-modulus CFRP plates and steel. It was revealed that the influence is minimal (less than 4.5%). The effect of fatigue loading on stiffness is less than 10%. A local fatigue damage zone near the joint of double-shear pull test specimen was observed by Wu et al. (2012b). The length of this local damage zone was found to be less than 1% of the bond length, which explains, to some extent, the reason why the influence of the fatigue cycle is not significant.

Miller et al. (2001) conducted fatigue tests on two full-scale bridge girders rehabilitated with CFRP plates for 10 million cycles at a stress range that might be expected in the field. Throughout the 10 million cycles, the CFRP plates were periodically monitored and inspected for debonding. Visual inspections and tapping tests were unable to uncover any evidence of debonding due to the 10 million fatigue cycles. Furthermore, static testing revealed no change in global stiffness.

3.7 EFFECT OF IMPACT LOADING ON BOND STRENGTH

3.7.1 Effect of impact loading on material properties

Significant variation in the material overall mechanical properties is more likely to occur under impact loading in comparison with quasi-static ones. Al-Zubaidy et al. (2013) carried out an experimental investigation on the mechanical properties of the unidirectional normal-modulus CFRP sheet as well as the properties of Araldite 420 epoxy resin under medium-impact tensile loads. The strain rates for quasi-static tests are 2.42×10^{-4} s^{-1} and 6.66×10^{-4} s^{-1} on unidirectional normal-modulus CFRP and epoxy, respectively. The impact tests have strain rates of 54.2, 67.2, and 87.4 s^{-1} for both CFRP sheet and epoxy, which are about 80,000 to 360,000 times those of static tests.

The influence of strain rate on CFRP properties (tensile strength, modulus of elasticity, and strain at failure) is shown in Figure 3.18. There is about 20 to 40% increase in tensile strength. The increase in modulus of elasticity is about 20%, whereas the increase in strain at failure ranges from 7 to 24%.

The influence of strain rate on Araldite 420 properties (tensile strength, modulus of elasticity, and strain at failure) is also shown in Figure 3.18. The increase in tensile strength and modulus of elasticity of Araldite was much higher compared to CFRP. The impact tensile strength is about three times that of static tensile stress, and the impact modulus of elasticity is about twice that of static values. However, the strain at failure of adhesive reduces about 50% under such impact loading. This may be due to the fact that adhesive is more brittle than CFRP.

3.7.2 Effect of impact loading on bond strength

Al-Zubaidy et al. (2012) conducted some double-shear pull tests to investigate the bond characteristics between CFRP sheets and steel plates under impact tensile loads. Araldite 420 was used as the adhesive in the testing program. The bond length varied from 10 to 100 mm. The number of CFRP layers was taken as one and three. The loading rate is around 2 mm/min for static test and 4.26 m/s for impact test. Figure 3.19 shows the test setup, where a special rig was built to convert the impact load to tensile load on the specimen.

The effective bond length defined in Section 3.4.3 was found to be around 40 mm under impact loading, whereas the effective bong length is about 50 mm under static loading.

Six failure modes were illustrated for CFRP–steel systems in Figure 3.3. When the bond length is less than the effective bond length, the failure mode changes from mode 1 (i.e., adhesion failure between steel and adhesive) under static load to mode 4 (i.e., CFRP delamination: separation of some carbon fibres from the resin matrix) under dynamic load. When the bond length is greater than the effective bond length and only one layer of CFRP is applied, the failure mode changes from mode 5 (i.e., CFRP rupture) under static load to mode 5 and mode 4. When three layers of CFRP are applied, the failure mode changes from combined modes 1 and 4 to mode 4.

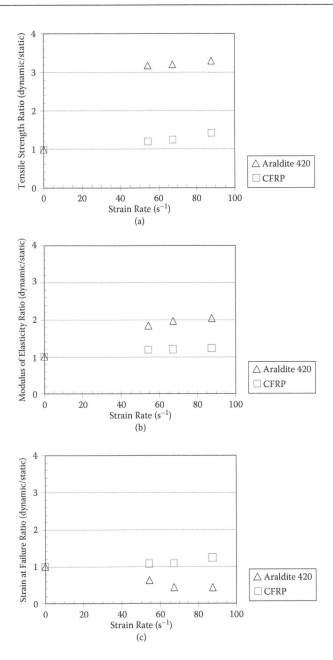

Figure 3.18 Effect of strain rate on properties of CFRP and Araldite. (a) Tensile stress. (b) Modulus of elasticity. (c) Strain to failure. (Adapted from Al-Zubaidy, H. et al., *Composite Structures*, 96(2), 153–164, 2013.)

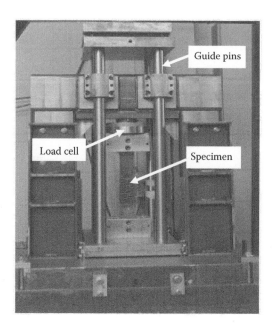

Figure 3.19 Impact test setup. (Courtesy of H. Al-Zubaidy, Monash University, Australia.)

The effect of strain rate on bond strength is shown in Figure 3.20. It can be seen that more influence is observed for three layers of the CFRP–steel system with an average increase of 70%.

3.8 DURABILITY OF BOND BETWEEN FRP AND METAL

In engineering terms, durability of a material or structure means the ability to resist cracking, oxidation, chemical degradation, delamination, and wear for a specific period of time, under the appropriate load conditions and specific environmental conditions (Hollaway and Cadei 2002). There are a few reviews (Karbhari et al. 2003, Hollaway 2010) on the importance and complexities of the durability problem and the potential opportunities of research.

Baker et al. (2002) presented a comprehensive summary of bonded composite repair of metallic aircraft structure. The book *Durability of Composites for Civil Structural Applications*, edited by Karbhari (2003), provides very detailed insight into the durability of fibre-reinforced composites under all environmental and physical conditions. These areas include fibre-reinforced composites under aqueous solution, thermal effects, fatigue loading, creep, and fire. The book also discusses material

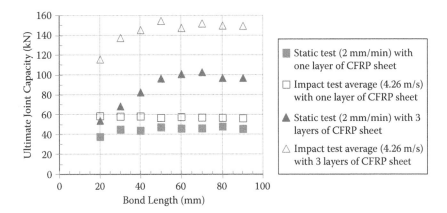

Figure 3.20 Effect of strain rate on bond strength. (Adapted from Al-Zubaidy, H. et al., *Composite Structures*, 94(11), 3258–3270, 2012.)

properties, resin, and adhesive types suitable for civil application and manufacturing processes.

Extensive research has been conducted on durability of the bond between CFRP and concrete. For example, Wan et al. (2006) showed that the presence of water during primer application decreases the bond quality significantly. Water presence also changes the mode of failure, from cohesive through concrete to adhesive along the concrete-primer interface. It was also found that moisture causes deterioration of the bond after the epoxy has cured. The percentage of degradation was dependent on the time of exposure. Dai et al. (2010) investigated the long-term durability of adhesives subject to wet-dry cycles. All FRP-to-concrete bonded joints failed at the interface between the primer and concrete after exposure, while those not exposed usually failed within the concrete substrate. Tuakta and Büyüköztürk (2011) found that degradation of bond strength could be up to 70% after just 8 weeks of moisture conditioning. Karbhari and Navada (2008) showed that surface preparation has a significant effect on both fracture energy release rate and rate of crack propagation, which can be accelerated by exposure.

The knowledge of repairing aircraft cannot be directly applied to civil structural applications for several reasons: (1) Aircraft is made of aluminium and the composites used for repairing are the boron-epoxy and graphite-epoxy systems. (2) Boron fibre is too expensive for civil engineering applications. (3) The modulus of elasticity for graphite-epoxy composites is about 148 GPa, which is rather low for strengthening steel structures with a modulus of elasticity of 200 GPa. The CFRP to be commonly used has a modulus of elasticity (E_{CFRP}) ranging from 240 to 640 GPa. (4) The failure modes of CFRP–steel systems are different from those observed in aluminium aircraft structures repaired with boron- or graphite-epoxy systems. The knowledge of CFRP-concrete systems cannot be directly applied to CFRP–steel systems because of different failure modes in the two systems, as explained in Section 1.3.

A schematic view of the CFRP–steel system under environmental conditions is shown in Figure 3.21. Research on the combined role of corrosive environment and mechanical loading in material degradation and mechanical properties of CFRP-strengthened steels has been limited. Two recent studies are summarised below.

Dawood and Rizkalla (2010) conducted 44 tests on CFRP plate–steel double-shear pull test joints exposed to severe environmental conditions for different durations, up to 6 months. Specimens that were bonded using a thin layer of adhesive exhibited a 60% degradation of the measured bond strength after 6 months of exposure, with some specimens debonding spontaneously during the exposure period. Specimens that were pretreated with a silane coupling agent prior to bonding exhibited essentially no degradation of the bond strength over the 6-month exposure duration. This research only addressed CFRP plate–steel double-shear pull test joints under wet-dry cycling in salt water at one constant temperature (38°C). The CFRP–steel system applied using the wet lay-up method was not examined, where the environmental effect may be more severe without a clear adhesive layer as an insulator between the CFRP and steel.

Nguyen et al. (2012) carried out a series of double-shear pull tests on CFRP sheet-steel joints at different load levels (i.e., 80, 50, and 20% of their

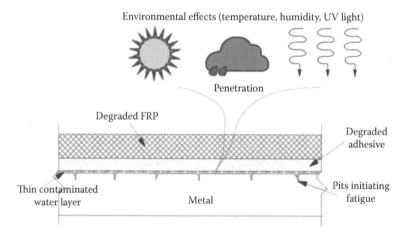

Figure 3.21 CFRP–steel system under environmental conditions. (Adapted from Zhao, X.L. et al., Durability of Carbon Fibre Reinforced Polymer (CFRP) Strengthened Steel Structures against Environment-Assisted Degradation, Australian Research Council (ARC) Discovery Grant Application for Funding in 2012 to 2014, awarded in November 2011.)

ultimate load measured at room temperature) with constant temperatures from 35 to 50°C (i.e., temperatures below and above the glass transition temperature (T_g, 42°C) of the adhesive) or with cyclic thermal loading between 20 and 50°C. It was found that adhesively bonded steel-CFRP double-shear pull test joints exhibit an obvious time-dependent behaviour, i.e., higher target temperature or larger applied load results in shorter time to failure. It was recommended that in practice the adhesive layer should be kept at least 7 to 10°C below T_g of the adhesive to avoid strength degradation of the structure due to temperature effect.

3.9 FUTURE WORK

Surface preparation is an important step in facilitating bond formation by providing a chemically active surface free of contaminants (Schnerch et al. 2006, Kim et al. 2010, Teng et al. 2010). It is the surface properties of the adherends, such as chemical composition and texture, that control adhesion failure and debonding (Teng et al. 2010). There is a need to quantify the surface condition and its relationship to the bond behaviour. Some preliminary work was reported in Fernando et al. (2013).

When the carbon fibres of the composite come into contact with steel, galvanic corrosion of steel may occur, coupled with blisters forming on the surface of the composite (Tavakkolizadeh and Saadatmanesh 2001, Torres-Acosta 2002). This may greatly impact the performance of the overall strengthened structure. Attempt was made by researchers (e.g., Hollaway and Cadei 2002, Photiou et al. 2006, Shaat and Fam 2007) to reduce or prevent the galvanic corrosion by applying one layer of glass fibre-reinforced polymer (GFRP) before applying CFRP. There is a need to quantify the galvanic corrosion and its relationship to the reduction in bond strength.

Future research is needed to address the following critical issues regarding durability:

1. The degradation in the chemical nature of CFRP and adhesive, due to environmental factors, i.e., chloride, ions, UV radiation, humidity, and temperature
2. As a result of (1) above, degradation in the mechanical properties (under tensile and cyclic loading) of CFRP and adhesive, due to the same environmental factors
3. Condensation of moisture and enrichment of chloride ions at the adhesive-steel interface, which produce an ideal condition for pitting and crevices:
   a. Pits are common initiators of fatigue and stress corrosion cracks.
   b. Corrosion of steel at the adhesive-steel interface may cause delamination of adhesive from the steel substrate.

4. The combined effect of environmental factors and mechanical loading on the mechanical and material properties of individual components in the CFRP–steel system, i.e., CFRP plate–sheet, adhesive and steel

5. The combined effect of environmental factors and mechanical loading on the bond between CFRP and steel

More work is needed to understand the influence of strain rate on bond between FRP and metal since there is a potential to use FRP to strengthen metallic structures subject to impact and blast loading.

REFERENCES

Al-Shawaf, A., Al-Mahaidi, R., and Zhao, X.L. 2005. Tensile properties of CFRP laminates at subzero temperatures. In *Proceedings of the Australian Structural Engineering Conference*, Newcastle, UK, September 11–14, pp. 1–9.

Al-Shawaf, A., Al-Mahaidi, R., and Zhao, X.L. 2006. Study on bond characteristics of CFRP/steel double-lap shear joints at subzero temperature exposure. In *Proceedings of the Third International Conference on FRP Composites in Civil Engineering (CICE 2006)*, Miami, December 13–15, pp. 71–74.

Al-Shawaf, A., and Zhao, X.L. 2013. Adhesive rheology impact on wet layup CFRP/steel joints' behaviour under infrastructural subzero exposures. *Composites Part B: Engineering*, 47, 207–219.

Al-Zubaidy, H., Al-Mahaidi, R., and Zhao, X.L. 2012. Dynamic bond strength between CFRP sheet and steel. *Composite Structures*, 94(11), 3258–3270.

Al-Zubaidy, H., Zhao, X.L., and Al-Mahaidi, R. 2013. Mechanical characterization of the dynamic tensile properties of CFRP sheet and adhesive at medium strain rates. *Composite Structures*, 96(2), 153–164.

Bai, Y., and Keller, T. 2011. Effects of thermal loading history on structural adhesive modulus across glass transition. *Construction and Building Materials*, 25(4), 2162–2168.

Bai, Y., Keller, T., and Vallee, T. 2008. Modeling of stiffness of FRP composites under elevated and high temperature. *Composites Science and Technology*, 68(15–16), 3099–3106.

Baker, A.A., Rose, L.R.F., and Jones, R. 2002. *Advances in the bonded composite repair of metallic aircraft structure*. Amsterdam: Elsevier.

Colombi, P., and Poggi, C. 2006. Strengthening of tensile steel members and bolted joints using adhesively bonded CFRP plates. *Construction and Building Materials*, 20(1–2), 22–33.

Dai, J.G., Yokota, H., Iwanami, H., and Kato, E. 2010. Experimental investigation of the influence of moisture on the bond behaviour of FRP to concrete interfaces. *Journal of Composites for Construction*, 14(6), 834–844.

Dawood, M., and Rizkalla, S. 2010. Environmental durability of a CFRP system for strengthening steel structures. *Construction and Building Materials*, 24(9), 1682–1689.

El Damatty, A.A., and Abushagur, M. 2003. Testing and modelling of shear and peel behaviour for bonded steel/FRP connections. *Thin-Walled Structures*, 41(11), 987–1003.

Fawzia, S. 2007. Bond characteristics between steel and carbon fibre reinforced polymer (CFRP) composites. PhD thesis, Department of Civil Engineering, Monash University, Melbourne.

Fawzia, S., Al-Mahaidi, R., and Zhao, X.L. 2006. Experimental and finite element analysis of a double strap joint between steel plates and normal modulus CFRP. *Composite Structures*, 75(1–4), 156–162.

Fawzia, S., Al-Mahaidi, R., Zhao, X.L., and Rizkalla, S. 2007. Strengthening of circular hollow steel tubular sections using high modulus CFRP sheets. *Construction and Building Materials*, 21(4), 839–847.

Fawzia, S., Zhao, X.L., and Al-Mahaidi, R. 2010. Bond–slip models for double strap joints strengthened by CFRP. *Composite Structures*, 92(9), 2137–2145.

Fernando, D., Teng, J.G., Yu, T., and Zhao, X.L. 2013. Preparation and characterization of steel surface for adhesive bonding. *Journal of Composites for Construction*, in press.

Hart-Smith, L.J. 1973. *Adhesive-bonded double-lap joints*. Technical Report NASA CR-112235. Long Beach, CA: Douglas Aircraft Company.

Hollaway, L.C. 2010. A review of the present and future utilisation of FRP composites in the civil infrastructure with reference to their important in-service properties. *Construction and Building Materials*, 24(12), 2419–2445.

Hollaway, L.C., and Cadei, J. 2002. Progress in the technique of upgrading metallic structures with advanced polymer composites. *Progress in Structural Engineering and Materials*, 4(2), 131–148.

Hollaway, L.C., and Teng, J.G. 2008. *Strengthening and rehabilitation of civil infrastructures using fibre-reinforced polymer (FRP) composites*. Cambridge, England: Woodhead Publishing Limited.

Jiao, H., and Zhao, X.L. 2004. CFRP strengthened butt-welded very high strength (VHS) circular steel tubes. *Thin-Walled Structures*, 42(7), 963–978.

Karbhari, V.M. 2003. *Durability of composites for civil structural applications*. Oxford: Woodhead Publishing Limited.

Karbhari, V.M., Chin, J.W., Hunston, D., Benmokrane, B., Juska, T., Morgan, R., Lesko, J.J., Sorathia, U., and Reynaud, D. 2003. Durability gap analysis for fiber-reinforced polymer composites in civil infrastructure. *Journal of Composites in Construction*, 7(3), 238–247.

Karbhari, V.M., and Navada, R. 2008. Investigation of durability and surface preparation associated defect criticality of composites bonded to concrete. *Composites Part A: Applied Science and Manufacturing*, 39(6), 997–1006.

Kim, S.J., Smith, S.T., and Young, B. 2010. Effect of surface preparation on the strength of FRP-to-mild steel and FRP-to-stainless steel joints. Presented at *Proceedings of the 5th International Conference on Composites in Civil Engineering*, Beijing, September 27–29.

Liu, H.B., Zhao, X.L., and Al-Mahaidi, R. 2010. Effect of fatigue loading on bond strength between CFRP sheets and steel plates. *International Journal of Structural Stability and Dynamics*, 10(1), 1–20.

Liu, H.B., Zhao, X.L., Al-Mahaidi, R., and Rizkalla, S.H. 2005. Analytical bond models between steel and normal modulus CFRP. In *Proceedings of Fourth International Conference on Advances in Steel Structures*, Shanghai, China, June 13–15, pp. 1545–1552.

Matta, F. 2003. Bond between steel and CFRP laminates for rehabilitation of metallic bridges. Master degree thesis, Faculty of Engineering, University of Padua, Padua.

Miller, T.C., Chajes, M.J., Mertz, D.R., and Hastings, J.N. 2001. Strengthening of a steel bridge girder using CFRP plates. *Journal of Bridge Engineering*, 6(6), 514–522.

Nguyen, T.C., Bai, Y., Zhao, X.L., and Al-Mahaidi, R. 2011. Mechanical characterization of steel/CFRP double strap joints at elevated temperatures. *Composite Structures*, 93(6), 1604–1612.

Nguyen, T., Yu, B., Al-Mahaidi, R., and Zhao, X.L. 2012. Time-dependent behaviour of steel/CFRP double strap joints subjected to combined thermal and mechanical loading. *Composite Structures*, 94(5), 1826–1833.

Pham, H.B., and Al-Mahaidi, R. 2005. Modelling of CFRP-concrete shear-lap tests. *Construction and Building Materials*, 21(4), 727–735.

Photiou, N.K., Hollaway, L.C., and Chryssanthopoulos, M.K. 2006. Strengthening of an artificially degraded steel beam utilizing a carbon/glass composite system. *Construction and Building Materials*, 20(1–2), 11–21.

Sauder, C., Lamon, J., and Pailler, R. 2004. The tensile behavior of carbon fibers at high temperatures up to 2400°C. *Carbon*, 42(4), 715–725.

Schnerch, D., Dawood, M., Rizkalla, S., Sumner, E., and Stanford, K. 2006. Bond behavior of CFRP strengthened steel structures. *Advances in Structural Engineering—An International Journal*, 9(6), 805–817.

Schnerch, D., Stanford, K., Summer, E., and Rizkalla, S. 2004. *Strengthening steel structures and bridges with high modulus carbon fiber reinforced polymers: Resin selection and scaled monopole behaviour*, pp. 237–245. Transportation Research Record 1892.

Shaat, A., and Fam, A. 2007. Fiber-element model for slender HSS columns retrofitted with bonded high-modulus composites. *Journal of Structural Engineering*, 133(1), 85–95.

Standards Australia. 1998. *Steel structures*. Australian Standard AS 4100. Sydney: Standards Australia.

Tavakkolizadeh, M., and Saadatmanesh, H. 2001. Galvanic corrosion of carbon and steel in aggressive environments. *Journal of Composites for Construction*, 5(3), 200–210.

Teng, J.G., Fernando, D., Yu, T., and Zhao, X.L. 2010. Treatment of steel surfaces for effective adhesive bonding. Presented at *Proceedings of the 5th International Conference on Composites in Civil Engineering*, Beijing, September 27–29.

Teng, J.G., Fernando, D., Yu, T., and Zhao, X.L. 2012. Debonding failures in CFRP-strengthened steel structures. Presented at Third Asia-Pacific Conference on FRP in Structures (APFIS2012), Sapporo, Japan, February 2–4.

Torres-Acosta, A.A. 2002. Galvanic corrosion of steel in contact with carbon-polymer composites. II. Experiments in concrete. *Journal of Composites for Construction*, 6(2), 116–122.

Tuakta, C., and Büyüköztürk, O. 2011. Deterioration of FRP/concrete bond system under variable moisture conditions quantified by fracture mechanics. *Composites Part B: Engineering*, 42(2), 145–154.

Wan, B., Petrou, M.F., and Harries, K.A. 2006. The effect of the presence of water on the durability of bond between CFRP and concrete. *Journal of Reinforced Plastics and Composites*, 25(8), 875–890.

Wu, C., Zhao, X.L., Chiu, W.K., Al-Mahaidi, R., and Duan, W.H. 2012b. Effect of fatigue loading on the bond behaviour between UHM CFRP plates and steel plates. *Composites Part B: Engineering*, accepted for publication.

Wu, C., Zhao, X.L., Duan, W.H., and Al-Mahaidi, R. 2012a. Bond characteristics between ultra high modulus CFRP laminates and steel. *Thin-Walled Structures*, 51(2), 147–157.

Wu, Z.S., Yuan, H., and Niu, H.D. 2002. Stress transfer and fracture propagation in different kinds of adhesive joints. *Journal of Engineering Mechanics*, 128(5), 562–573.

Xia, S.H., and Teng, J.G. 2005. Behavior of FRP-to-steel bond joints. In *Proceedings of International Symposium on Bond Behaviour of FRP in Structures (BBFS 2005)*, Hong Kong, December, pp. 419–426.

Yao, J., Teng, J.G., and Chen, J.F. 2005. Experimental study on FRP-to-concrete bonded joints. *Composites Part B: Engineering*, 36(12), 99–113.

Yu, T., Fernando, D., Teng, J.G., and Zhao, X.L. 2012. Experimental study on CFRP-to-steel bonded interfaces. *Composites Part B: Engineering*, 43(5), 2279–2289.

Yuan, H., Teng, J.G., Seracino, R., Wu, Z.S., and Yao, J. 2004. Full-range behaviour of FRP-to-concrete bonded joints. *Engineering Structures*, 26(5), 553–565.

Zhao, X.L., Singh, R., Bai, Y., Bandyopadhyay, S., and Rizkalla, S. 2011. Durability of carbon fibre reinforced polymer (CFRP) strengthened steel structures against environment-assisted degradation. Australian Research Council (ARC) Discovery Grant Application for funding in 2012 to 2014, awarded in November 2011.

Zhao, X.L., and Zhang, L. 2007. State-of-the-art review on FRP strengthened steel structures. *Engineering Structures*, 29(8), 1808–1823.

Chapter 4

Flexural strengthening of steel and steel–concrete composite beams with FRP laminates

J. G. Teng

Department of Civil and Environmental Engineering,

Hong Kong Polytechnic University, Hong Kong, China

D. Fernando

School of Civil Engineering,

The University of Queensland, Queensland, Australia

4.1 GENERAL

This chapter is concerned with the flexural strengthening of steel and steel–concrete composite beams using adhesively bonded fibre-reinforced polymer (FRP) laminates. For brevity of presentation in this chapter, when referring to steel beams and steel–concrete composite beams collectively, the term *steel/composite beams* is used, and the word *composite* here should not be confused with the use of the same word to describe FRP as a composite material. For the same reason, the term *FRP laminates* is used to refer to both FRP sheets formed via the wet lay-up process and prefabricated FRP plates generally produced by the pultrusion process; the differentiation between FRP sheets and plates is only made when such differentiation is important. Nevertheless, *plate* is sometimes used in place of *laminate* in accordance with conventional terminology in construction-related research publications, particularly when the word *plate* is used to mainly indicate a geometrical feature (e.g., *FRP-plated beams* refers to beams strengthened with adhesively bonded FRP laminates); in such cases, differentiation between FRP sheets and plates is unimportant. Finally, in most cases of steel/composite beam strengthening, carbon fibre-reinforced polymer (CFRP) is the material of choice, but the term *FRP* is generally used in this chapter as the more inclusive term unless the use of CFRP is being emphasised.

The flexural strengthening of steel/composite beams has become a topic of substantial interest in recent years mainly due to the ageing stock of bridges that contain steel or steel–concrete composite girders. Traditional methods for the strengthening of steel/composite beams involve the attachment of steel plates or sections through welding, bolting, or adhesive bonding. In addition, the peening technique has recently been introduced to improve the fatigue life of welded steel joints (Maddox 1985). The addition of heavy steel plates or sections by welding or bolting, however, has several disadvantages compared with the adhesive bonding of lightweight FRP laminates: (1) the use of heavy steel plates or sections involves a labour-intensive and time-consuming installation procedure; (2) when bolting is used, hole drilling may lead to temporary/local weakening of the structure; (3) when welding is used, risks of weld fatigue cracking are introduced, and in addition, welding is not allowed in fire-sensitive environments; and (4) the significant weight of the added components may reduce the member's capacity to carry live loads and increase its deflections. Steel plates or sections may also be adhesively bonded to existing steel beams, in which case the risks of temporary/local weakening and fatigue cracking are avoided, but the high bending stiffness of steel plates/sections, compared to thin strong FRP laminates, makes premature debonding failure a more likely event. The peening technique is effective in the repair of small fatigue cracks, but quickly becomes ineffective as the crack depth increases; for cracks deeper than 2 mm, peening has little effect (Ghahremani and Walbridge 2011).

Due to the issues mentioned above with the addition of steel plates or sections, the adhesive bonding of lightweight strong FRP laminates has become an attractive method in recent years for the flexural strengthening of steel/composite beams in many practical situations. Apart from the high strength-to-weight ratio of FRP composites, FRP laminates also possess some other advantages over steel plates or sections in the strengthening of steel/composite beams, including (1) excellent corrosion resistance, (2) shape flexibility, and (3) directional tailorability of material properties. The first property may not be so important unless the original steel/composite beam is already well protected against corrosion, but the second and third properties are important in a variety of situations (e.g., strengthening of curved steel/composite beams with FRP laminates whose fibres are oriented only or mainly in the axial direction). Due to the advantages discussed above, adhesive bonding of FRP laminates can offer cost-effective flexural strengthening solutions for steel/composite beams despite their relatively high material cost.

Many studies have recently been conducted on the strengthening of steel/composite beams by the bonding of an FRP laminate to the soffit (i.e., tension flange of a simply supported beam) (e.g., Miller et al. 2001, Tavakkolizadeh and Saadatmanesh 2003a, Al-Saidy et al. 2004, Nozaka et al. 2005, Colombi and Poggi 2006, Lenwari et al. 2006, Sallam et al. 2006, Schnerch and Rizkalla 2008, Shaat and Fam 2008, Fam et al. 2009, Linghoff et al. 2009). These studies have shown that significant strength gains and, in some cases, significant stiffness gains can be achieved using adhesively bonded FRP laminates. Tavakkolizadeh and Saadatmanesh (2003a) reported 44 to 76% increases in the ultimate load of steel–concrete composite beams strengthened with epoxy-bonded CFRP sheets. Al-Saidy et al. (2004) reported increases of up to 45% in the ultimate load of steel–concrete composite beams due to epoxy-bonded CFRP plates. Schnerch and Rizkalla (2008) reported 10 to 34% stiffness increases and up to 46% strength increases for steel–concrete composite beams strengthened with ultra-high-modulus CFRP plates.

Both the strength gain and the stiffness gain, however, depend strongly on the type and amount of the FRP laminate. Typically CFRP is commonly preferred for use in the strengthening of steel/composite beams over other FRP materials due to its higher elastic modulus. The use of high-modulus CFRP is desirable when stiffness gains are needed to ensure either deflection reductions or buckling load increases.

The effectiveness of FRP flexural strengthening of steel beams without adequate lateral support depends also on the susceptibility of the section form to flexural–torsional (lateral buckling) failure. The most common steel section form for steel/composite beams is probably the hot-rolled I-section, which is representative of hot-rolled open sections. The rectangular hollow section (RHS) is another common section form for steel beams; RHS is representative of closed sections with a large torsional stiffness. An important difference between I-section beams and RHS beams is that the former may fail by lateral buckling, which is very unlikely for RHS beams. Therefore, the design theory for FRP-plated RHS beams may be seen as a subset of the design theory for FRP-plated I-section beams. For this reason, although the content of this chapter is largely applicable to both I-section and RHS steel/composite beams (i.e., except the parts on lateral and local buckling), this chapter is explicitly concerned only with I-section beams.

Apart from I-section and RHS steel beams, the strengthening of circular hollow section (CHS) steel beams using FRP sheets has also been explored (Seica and Packer 2007, Haedir and Zhao 2012). Due to local buckling failures and the important role played by hoop fibres of the FRP laminate used, the design of such FRP-strengthened CHS beams is significantly different from that of FRP-strengthened I-section or RHS beams as described in this chapter. Therefore, the design of such FRP-strengthened CHS beams is not addressed in this chapter. Readers should refer to Seica and Packer (2007) and Haedir and Zhao (2012) for further information on the topic.

In addition to the section form, the steel section slenderness also has a significant bearing on the effectiveness of FRP flexural strengthening of steel/composite beams. More specifically, for slender steel sections, the effectiveness of FRP strengthening can be substantially compromised by possible local buckling failures (e.g., compression flange buckling or web buckling). For this reason, this chapter is only concerned with steel/composite beams with a compact steel section for which local buckling does not occur until the section has undergone a significant amount of plastic deformation. It should be noted that local buckling may become a more likely or more significant event as a result of FRP strengthening, which can lead to greater compressive straining of the steel section.

This chapter therefore aims to present a systematic treatment of the behaviour and design of steel and steel–concrete composite beams of compact steel I-sections strengthened with FRP laminates bonded to the tension flange; the information presented in the chapter is also largely applicable to steel/composite beams of other compact steel sections, except when lateral or local buckling is of concern. All discussions are concerned with only simply supported beams, which have been the predominant concern of existing research unless explicitly specified otherwise. While FRP laminates may be bonded to other parts of a beam instead of the tension flange for flexural strengthening purposes, such options are not considered in this chapter, as there has been little research in the published literature on such options. In some strengthening applications, it may be beneficial to prestress the FRP laminate before adhesive bonding, but this aspect is also not considered due to the same reason of inadequacy of existing research.

The chapter starts with explanations of different failure modes, followed by descriptions of theoretical strength models for these failure modes. A rational design procedure that addresses the different failure modes in an integrated manner is then presented and demonstrated using a design example. The chapter ends with discussions of future research needs. The treatment in this chapter of the behaviour and design of FRP-plated steel–concrete composite beams is based on the work presented in Fernando and Teng (2013). Readers are referred to Fernando and Teng (2013) for more details.

4.2 FAILURE MODES

4.2.1 General

Provided that local buckling does not occur, the strength of FRP-plated steel/composite beams is governed by either in-plane failure or lateral buckling. The in-plane failure modes of FRP-plated steel/composite beams include (a) in-plane bending failure due to the tensile rupture of FRP or the compressive crushing of concrete, with the latter mode being applicable only to steel–concrete composite beams; (b) debonding at the ends of the FRP laminate (referred to as end debonding); and (c) cracking or yielding-induced debonding between the two ends of the FRP laminate (referred to as intermediate debonding). The latter two failure modes are similar to end debonding and intermediate crack debonding in FRP-plated concrete beams. In addition, FRP-plated I-section steel beams without adequate lateral support (e.g., from a concrete slab) may suffer out-of-plane failure by lateral buckling (failure mode d).

Even though this chapter is limited to the strengthening of beams with a compact steel section, it needs to be pointed out that local buckling of plate elements still needs to be considered in design. This is because the addition of an FRP laminate leads to a shift of the neutral axis toward the tension flange, leading to the development of greater compressive strains in the flange and the web at the attainment of the ultimate load. As a result of the neutral axis shift, the following local buckling modes may occur: (e) local buckling of the compression flange and (f) local buckling of the web.

4.2.2 In-plane bending failure

In-plane bending failure of a steel/composite beam with an FRP laminate bonded to the tension flange occurs when the strain of the extreme tension or compression fibre of the section reaches its limiting value. For FRP-plated steel beams, it is easy to understand that in-plane bending failure occurs when the strain of the extreme tension fibre in the FRP laminate reaches its rupture strain. For FRP-plated steel–concrete composite beams, in-plane bending failure is often also governed by the tensile rupture of the FRP laminate, especially when a high-modulus CFRP laminate with a relatively small rupture strain is used. However, if a normal-modulus CFRP laminate is used, failure is often due to concrete crushing.

A typical in-plane bending failure of an FRP-plated steel–concrete composite beam, due to the tensile rupture of the FRP laminate, is shown in Figure 4.1; the load-displacement curve of this beam is shown in Figure 4.2. For an FRP-plated steel/composite beam failing by FRP rupture, a sudden drop in the load resisted by the beam can be expected at the instant of FRP

Figure 4.1 In-plane bending failure of FRP-plated steel–concrete composite beam by FRP rupture. (Courtesy of Drs. D. Schnerch and S. Rizkalla.)

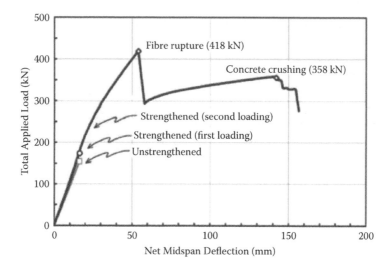

Figure 4.2 Typical load-displacement curve of FRP-plated steel–concrete composite beam failing by FRP rupture. (Courtesy of Drs. D. Schnerch and S. Rizkalla.)

rupture. However, the load drop is limited by the residual capacity of the plated beam (Figure 4.2), which is generally the same as the load carrying capacity of the original unstrengthened beam. If compressive crushing of concrete governs the strength, then the load drop is generally more drastic and the residual capacity more limited; the process of failure is also brittle, like that of overreinforced concrete beams.

4.2.3 Lateral buckling

Steel beams may fail in the lateral buckling mode, in which the member suddenly deflects laterally and twists out of the plane of loading. Lateral buckling occurs when the beam is weak in resisting lateral bending and torsional deformations; it is thus likely for open-section beams such as I-section beams without adequate lateral support, especially when the beam is slender, but is very unlikely for closed-section beams such as RHS beams. Obviously, lateral buckling cannot occur in a steel–concrete composite beam due to the presence of the concrete slab. The lateral buckling failure of an I-section steel beam is shown in Figure 4.3.

The tendency of a steel beam to buckle laterally can be significantly increased as a result of flexural strengthening by a bonded tension face FRP laminate. This is because such strengthening increases the in-plane bending strength much more effectively than the lateral buckling strength, although the latter may also be increased significantly. It is therefore important, in the design of an FRP strengthening system, to check whether lateral

Figure 4.3 Lateral buckling of I-section steel beam. (Reprinted from Trahair, N.S., *Flexural–Torsional Buckling of Structures*, E & FN Spon, London, 1993.)

buckling will become the governing failure mode of an FRP-plated steel beam, which is governed by in-plane bending failure before strengthening. The in-plane bending strength of such an FRP-plated steel beam can be easily determined, provided that debonding does not become a critical failure mode, but the evaluation of its lateral buckling strength is less straightforward.

4.2.4 End debonding

Of the two debonding failure modes, end debonding has received more attention (Sen et al. 2001, Deng and Lee 2007a, Schnerch et al. 2007). End debonding of FRP-plated steel/composite beams occurs due to high localised interfacial peeling stresses and shear stresses near the laminate end, and these high interfacial stresses arise due to the need to transfer stresses in the laminate to the beam near the laminate end. Typically, end debonding initiates at a laminate end and then propagates quickly toward the higher moment region, exhibiting the characteristics of a brittle, unstable process. The end debonding failure of an FRP-plated steel beam is shown in Figure 4.4.

Figure 4.4 End debonding failure. (Courtesy of Drs. M. Dawood, M. Guddati, and S. Rizkalla.)

4.2.5 Intermediate debonding

Another possible debonding failure mode in FRP-plated steel/composite beams is referred to as intermediate debonding, which initiates due to the presence of localised damage (e.g., cracking or localised yielding of steel) (Sallam et al. 2006) in a high-moment region (Figure 4.5). Intermediate debonding thus generally initiates in a region where the FRP laminate is highly stressed, and moves toward one of the laminate ends where the axial stress in the FRP laminate is low. Compared with end debonding, intermediate debonding is generally less sudden, but it is still a brittle failure process. Very limited research is available on intermediate debonding in FRP-plated steel/composite beams (Sallam et al. 2006).

Intermediate debonding in FRP-plated steel/composite beams is similar to intermediate crack (IC) debonding in FRP-plated concrete beams (Teng et al. 2003), in the sense that both initiate where the substrate is locally damaged/weakened and the interface is consequently subjected to high localised interfacial shear stresses. Therefore, it can be expected that the intermediate debonding strength depends strongly on the interfacial shear (i.e., mode II) fracture energy obtained from simple bonded joint tests (e.g., from single-lap shear tests as described in Yu et al. (2012)).

4.2.6 Local buckling of plate elements

The load carrying capacity of a steel/composite beam with a compact steel section may be substantially increased by the bonding of an FRP laminate to its tension flange. As a result of FRP strengthening, a larger

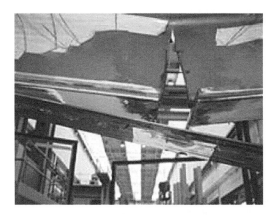

Figure 4.5 Intermediate debonding in a cracked steel beam.

Figure 4.6 Compression flange buckling of an unstrengthened steel I-beam. (Reprinted from Green, S. et al., *Journal of Constructional Steel Research*, 58(5–8), 907–941, 2002.)

portion of the cross section is placed in compression (i.e., a shift of the neutral axis toward the tension flange is induced). Consequently, the compression flange and the web are subjected to higher compressive strains under the increased ultimate load. Therefore, even though in the unstrengthened beam compression flange or web buckling cannot occur, the FRP-plated beam may suffer compression flange or web buckling. A typical compression flange buckling failure of a steel beam is shown in Figure 4.6. This failure mode can be avoided by strengthening the compression flange with bonded FRP laminates. A typical web buckling failure in a steel–concrete composite beam is shown in Figure 4.7. The use of bonded

Figure 4.7 Web buckling of an unstrengthened steel–concrete composite beam in negative moment region. (Reprinted from Vasdravellis, G. et al., *Journal of Constructional Steel Research*, 79, 34–47, 2012.)

FRP laminates to strengthen the web against local buckling has also been explored (Sayed-Ahmed 2006).

4.3 FLEXURAL CAPACITY OF FRP-PLATED STEEL/COMPOSITE SECTIONS

4.3.1 General

Provided buckling and debonding failures do not occur, the flexural strength of an FRP-plated steel/composite beam is governed by the (full) flexural capacity of the plated section at either FRP tensile rupture or concrete crushing. This is the most desirable failure mode for an FRP-plated steel/composite beam, and its corresponding section capacity can be predicted by a conventional section analysis. Such section analysis has been given in design guidelines for both FRP-plated steel beams (CNR-DT 2005, ICE 2001) and FRP-plated steel–concrete composite beams (Schnerch et al. 2007). In such a section analysis, the basic assumption made is that plane sections remain plane. The plane section assumption implies the assumption of full composite action between the original beam and the bonded FRP laminate, as well as between the steel section and the concrete slab.

In the following subsections, equations are presented for the section moment capacities of FRP-plated steel beams and FRP-plated steel–concrete composite beams separately for better clarity. In addition to the plane section assumption, the following assumptions are also made in developing the equations in the following subsections:

1. For steel–concrete composite beams, the tensile strength of concrete is ignored.
2. The steel has an elastic, perfectly plastic stress–strain relationship.

3. The FRP material behaves in a linear-elastic manner in tension until brittle tensile rupture occurs.

4.3.2 FRP-plated steel sections

The moment capacity of a compact steel beam section can be found without difficulty using a traditional plastic section analysis in which the section is assumed to have fully yielded. However, for an FRP-plated steel beam, it is, in principle, inappropriate to assume the development of a fully plastic section at the attainment of the ultimate moment. Instead, the central part of the steel section remains elastic and needs to be so treated in determining the ultimate moment. The assumption of a fully plastic section is valid only if most of the steel section yielded at the tensile rupture of the FRP laminate. On the basis of the assumptions described earlier for section analysis, the stress and strain distributions over the section depth of an FRP-plated steel section are shown in Figure 4.8. Compressive stresses and strains are taken to be positive. In Figure 4.8, h is the height of the original section, x is the neutral axis depth, b_{f1} and b_{f2} are the widths of the top and bottom flanges, respectively, b_{frp} is the width of the FRP laminate, t_{f1} and t_{f2} are the thicknesses of the top and bottom flanges, respectively, t_w is the web thickness, t_a is the thickness of the adhesive layer, t_{frp} is the thickness of the FRP laminate, $\varepsilon_{s,c}$ is the steel strain at the top surface of the top flange (i.e., compression face), $\varepsilon_{s,t}$ is the steel strain at the bottom surface of the bottom flange (i.e., tension face), $\varepsilon_{s,i}$ is the steel strain at a depth of h_i, ε_{frp} is the strain at the mid-surface of the FRP laminate, σ_{frp} is the average axial stress of the FRP laminate, $\sigma_{s,y}$ is the yield stress of steel, and $\sigma_{s,i}$ is the steel stress at a depth of h_i.

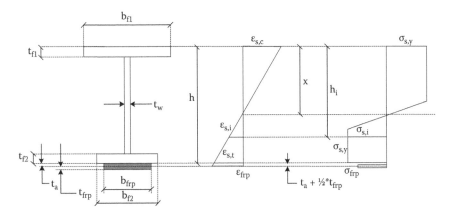

Figure 4.8 FRP-plated steel section and its strain and stress distributions.

The strains at the top and bottom surfaces of the steel beam (i.e., $\varepsilon_{s,c}$ and $\varepsilon_{s,t}$) can be related to the strain of the FRP laminate, ε_{frp}, as follows:

$$\varepsilon_{s,c} = \varepsilon_{frp} \frac{-x}{h + t_a + \dfrac{t_{frp}}{2} - x} \tag{4.1}$$

$$\varepsilon_{s,t} = \varepsilon_{frp} \frac{h - x}{h + t_a + \dfrac{t_{frp}}{2} - x} \tag{4.2}$$

$$\varepsilon_{s,i} = \varepsilon_{frp} \frac{h_i - x}{h + t_a + \dfrac{t_{frp}}{2} - x} \tag{4.3}$$

The depth of the neutral axis, x, can be determined by solving the following equilibrium equation:

$$\int_0^h \sigma_{s,i}\, b_i \, dh_i + \sigma_{frp} b_{frp} t_{frp} = 0 \tag{4.4}$$

where b_i is the width of the steel section at the depth of h_i. $\sigma_{s,i}$, b_i, and σ_{frp} are given by

$$\sigma_{s,i} = E_{steel}\varepsilon_{s,i} \quad \text{if} \quad \left|\varepsilon_{s,i}\right| < \frac{\sigma_{s,y}}{\gamma_s E_{steel}} \tag{4.5}$$

$$\sigma_{s,i} = \frac{\varepsilon_{s,i}}{\left|\varepsilon_{s,i}\right|} \frac{\sigma_{s,y}}{\gamma_s} \quad \text{if} \quad \left|\varepsilon_{s,i}\right| > \frac{\sigma_{s,y}}{\gamma_s E_{steel}} \tag{4.6}$$

$$b_i = b_{f1} \quad \text{if } 0 < h_i \le t_{f1} \tag{4.7}$$

$$b_i = t_w \quad \text{if } t_{f1} < h_i \le h - t_{f2} \tag{4.8}$$

$$b_i = b_{f2} \quad \text{if } h - t_{f2} < h_i \le h \tag{4.9}$$

and

$$\sigma_{frp} = E_{frp}\varepsilon_{frp} \le E_{frp}\varepsilon_{frpl,d} \tag{4.10}$$

where E_{steel} and E_{frp} are the moduli of elasticity of steel and FRP, respectively, γ_s is the partial safety factor for steel, and $\varepsilon_{frpl,d}$ is the design value of limiting strain of the FRP laminate.

The in-plane bending failure of such an FRP-plated steel beam is governed by the design value of FRP limiting strain, $\varepsilon_{frpl,d}$. Therefore, the design value of moment capacity of an FRP-plated steel beam can be obtained as

$$M_{ib,d} = \int_0^h \sigma_{s,i} b_i \left(\frac{h}{2} - h_i \right) dh_i + E_{frp} \varepsilon_{frpl,d} b_{frp} t_{frp} \left(\frac{h}{2} + t_a + \frac{t_{frp}}{2} \right) \quad (4.11)$$

If failure is due to the tensile rupture of FRP, then the design value of FRP limiting strain is equal to the design value of tensile rupture strain of FRP (i.e., $= \sigma_{frp,rup}/\gamma_{frp} E_{rup}$, in which $\sigma_{frp,rup}$ is the characteristic value of FRP tensile strength and γ_{frp} is the partial safety factor for FRP). Equation (4.11) can also be used to predict the design moment resistance at debonding failure by equating the design value of FRP limiting strain to the design value of debonding strain of the FRP laminate.

4.3.3 FRP-plated steel–concrete composite sections

The analysis of a steel–concrete composite section to determine its ultimate moment differs from that of an FRP-plated steel section in that the compressive force carried by the concrete section needs to be taken into account. In accordance with the plane section assumption, the axial strain varies linearly over the plated section (Figure 4.9). In Figure 4.9, b_e is the effective width of the concrete slab, h_c is the height (i.e., thickness) of the concrete slab, h is the height of the steel–concrete composite section, and $\varepsilon_{c,c}$ is the compressive strain at the top surface of the concrete slab.

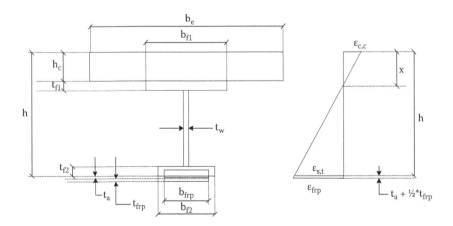

Figure 4.9 FRP-plated steel–concrete composite section and its strain distribution.

For an unstrengthened steel–concrete composite section, in-plane bending failure is governed by the compressive crushing of concrete, provided the shear connections between the steel and the concrete components do not fail. However, for an FRP-plated steel–concrete composite section, tensile rupture of FRP may occur before compressive crushing of concrete; therefore, in-plane bending failure may be due to either the tensile rupture of FRP or the compressive crushing of concrete. In the latter case, the traditional equivalent rectangular stress block for concrete can still be used, but the same stress block should not be used in the former case. This situation is similar to the section analysis of FRP-plated concrete sections (Teng et al. 2002a). Therefore, for the former case, the contribution of compressive stresses in the concrete to the moment capacity needs to be evaluated based on an appropriate stress–strain curve for concrete, such as that adopted by BS 8110 (1997) and more precisely defined in Kong and Evans (1987) (Figure 4.10).

If $\varepsilon_{c,c} = \varepsilon_{c,u}$ (ε_{cu} is the ultimate compressive strain of concrete) at the top surface of the concrete slab and $\varepsilon_{frp} = \varepsilon_{frpl,d}$ at the bottom surface of the FRP laminate when the ultimate moment of the section is reached, the section is referred to as a balanced section, and the critical depth of the neutral axis from the top surface of the concrete slab, x_c, can be written as

$$x_c = \frac{\varepsilon_{cu}}{\varepsilon_{cu} + \varepsilon_{frpl,d}} \left(h + t_a + \frac{t_{frp}}{2} \right) \qquad (4.12)$$

If the neutral axis depth $x > x_c$, failure occurs by compressive crushing of concrete, and if $x < x_c$, failure occurs by FRP tensile rupture. The neutral axis may be located in either the concrete slab or the steel beam. These two scenarios are both considered below.

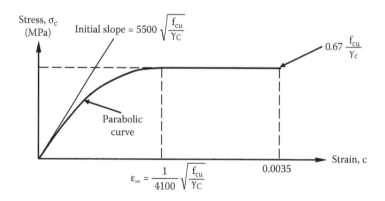

Figure 4.10 Stress–strain curve of concrete for design use adopted by BS 8110.

4.3.3.1 Neutral axis in the concrete slab

If the neutral axis is in the concrete slab, the strains and stresses over the section depth are as shown in Figure 4.11.

Based on the plane section assumption, the strain in the FRP, ε_{frp}, and the steel strain at $h_i > h_c$, $\varepsilon_{s,i}$ are related to the extreme compression fibre strain of concrete, $\varepsilon_{c,c}$, as follows:

$$\varepsilon_{frp} = -\varepsilon_{c,c}\frac{h + t_a + \dfrac{t_{frp}}{2} - x}{x} \tag{4.13}$$

$$\varepsilon_{s,i} = -\varepsilon_{c,c}\frac{h_i - x}{x} \tag{4.14}$$

According to the stress–strain curve shown in Figure 4.10, the compressive stresses in the concrete are given by (Kong and Evans 1987)

$$\sigma_c = 5500\left(\sqrt{\frac{f_{cu}}{\gamma_c}} - \frac{4100}{2}\varepsilon_c^2\right) \quad \text{if } 0 \le \varepsilon_c \le \varepsilon_{co} \tag{4.15}$$

and

$$\sigma_c = 0.67\frac{f_{cu}}{\gamma_c} \quad \text{if } \varepsilon_{co} \le \varepsilon_c \le 0.0035 \tag{4.16}$$

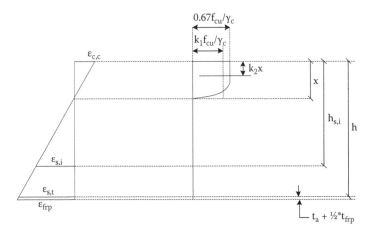

Figure 4.11 Strains and stresses over section depth when the neutral axis is within the concrete slab.

where σ_c is the compressive stress of concrete, and ε_{co} is the compressive strain of concrete at the first attainment of the peak axial stress, which has a value of

$$\varepsilon_{co} = \frac{1}{4100}\sqrt{\frac{f_{cu}}{\gamma_c}} \tag{4.17}$$

where f_{cu} is the characteristic cube compressive strength of concrete, and γ_c is the partial safety factor for concrete.

For any given ε_{co}, the total compression force of concrete, C, is given by

$$C = k_1 \frac{f_{cu}}{\gamma_c} b_e x \tag{4.18}$$

where k_1 is the mean stress factor defined by

$$k_1 = \frac{\displaystyle\int_0^{\varepsilon_{c,c}} \sigma_c \, d\varepsilon_c}{\left(f_{cu}/\gamma_c\right)\varepsilon_{c,c}} \tag{4.19}$$

The depth of the neutral axis x can be determined by solving the following force equilibrium equation:

$$k_1 \frac{f_{cu}}{\gamma_c} b_e x + \int_{h-h_c}^{h} \sigma_{s,i} b_i \, dh_i + \sigma_{frp} b_{frp} t_{frp} = 0 \tag{4.20}$$

$\sigma_{s,i}$ is given by Equations (4.5) and (4.6), σ_{frp} is given by Equation (4.10), and b_i is given by

$$b_i = b_{f1} \quad \text{if } h_c < h_i \le h_c + t_{f1} \tag{4.21}$$

$$b_i = t_w \quad \text{if } h_c + t_{f1} < h_i \le h - t_{f2} \tag{4.22}$$

$$b_i = b_{f2} \quad \text{if } h - t_{f2} < h_i \le h \tag{4.23}$$

The position of the compressive force of concrete is defined by D, which is the distance between the extreme concrete compression fibre to the line of action of the concrete compressive force. D can be related to the height of the compression zone through

$$D = k_2 x \tag{4.24}$$

where the centroidal factor k_2 of the compressive force is given by

$$k_2 = 1 - \frac{\int\limits_0^{\varepsilon_{c,c}} \varepsilon_c \sigma_c \, d\varepsilon_c}{\varepsilon_{c,c} \int\limits_0^{\varepsilon_{c,c}} \sigma_c \, d\varepsilon_c} \tag{4.25}$$

The design moment resistance of the beam is finally determined by

$$M_{ib,d} = k_1 \frac{f_{cu}}{\gamma_c} b_e x \left(\frac{h}{2} - k_2 x \right) + \int\limits_{h-h_c}^{h} \sigma_{s,i} b_i \left(\frac{h}{2} - h_i \right) dh_i$$

$$+ \sigma_{frp} b_{frp} t_{frp} \left(\frac{h}{2} + t_a + \frac{t_{frp}}{2} \right) \tag{4.26}$$

The beam is deemed to have reached failure when the concrete compressive strain attains the maximum useable strain (0.0035 according to BS 8110 (1997)) or the FRP reaches its design value of limiting strain.

If failure occurs by FRP tensile rupture, then it is possible that the strain of the extreme compression fibre of concrete, $\varepsilon_{c,c}$, is lower than ε_{co}, which is the compressive strain of concrete at the first attainment of the peak axial stress. In such cases, the stress distribution shown Figure 4.11 is no longer correct; the equations given above are, however, still applicable, but the constant stress region in the concrete block now becomes zero.

4.3.3.2 Neutral axis in the steel beam

If the neutral axis is in the top flange of the steel beam, the strains and stresses over the beam section are as shown in Figure 4.12.

When the neutral axis is in the steel beam, the entire concrete slab is in compression. Therefore, in addition to the values of ε_{cfrp} and $\varepsilon_{s,i}$, the strain at the bottom surface of the concrete slab, $\varepsilon_{c,b}$, also needs to be found. This strain can be related to the extreme compression fibre strain of concrete, $\varepsilon_{c,c}$, as follows:

$$\varepsilon_{c,b} = \varepsilon_{c,c} \frac{h_c}{x} \tag{4.27}$$

With the neutral axis in the steel beam, Equations (4.19) and (4.25) for evaluating the mean stress factor k_1 and the centroidal factor k_2 of the compressive force are no longer valid. The correct equations for k_1 and k_2 are as follows:

$$k_1 = \frac{\int\limits_{\varepsilon_{c,b}}^{\varepsilon_{c,c}} \sigma_c \, d\varepsilon_c}{\left(f_{cu}/\gamma_c \right) \varepsilon_{c,c}} \tag{4.28}$$

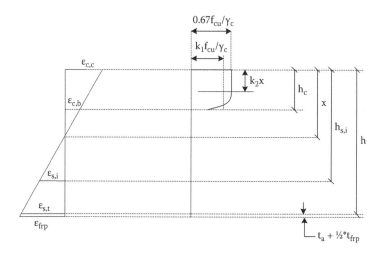

Figure 4.12 Strains and stresses over section depth when the neutral axis is within the steel beam.

$$k_2 = 1 - \dfrac{\displaystyle\int_{\varepsilon_{c,b}}^{\varepsilon_{c,c}} \varepsilon_c \sigma_c \, d\varepsilon_c}{\varepsilon_{c,c} \displaystyle\int_{\varepsilon_{c,b}}^{\varepsilon_{c,c}} \sigma_c \, d\varepsilon_c} \qquad (4.29)$$

Now by replacing Eqs. (4.19) and (4.25) with Eqs. (4.28) and (4.29), respectively, the equations given in Section 4.3.3.1 can still be used to calculate the design moment resistance of FRP-plated steel–concrete composite sections when the neutral axis is in the steel beam.

Again, if failure occurs by FRP tensile rupture and the strain of the extreme compression fibre of concrete, $\varepsilon_{c,c}$, is lower than ε_{co} (the compressive strain of concrete at the first attainment of peak axial stress), the equations given above are still applicable, but the constant stress portion of the concrete stress block no longer exists. That is, the stress distribution shown in Figure 4.12 is no longer correct.

4.3.4 Effects of preloading

The equations presented above are for beams without preloading. When strengthening is carried out on in-service beams, it is likely that some preloading (e.g., permanent dead loads) exists on the beam. The effect of preloading can be considered in strengthening design by modifying the strain value of the FRP laminate, as has been suggested for reinforced concrete (RC) beams strengthened with FRP laminates under preloading (Teng et al. 2002a). The FRP stress given by Equation (4.10) should be replaced with

$$
\sigma_{frp} =
\begin{cases}
0 & \text{if } \varepsilon_{frp} \leq \varepsilon_{t,i} \\
E_{frp}\left(\varepsilon_{frp} - \varepsilon_{t,i}\right) \leq E_{frp}\varepsilon_{frpl,d} & \text{if } \varepsilon_{frp} > \varepsilon_{t,i}
\end{cases}
\tag{4.30}
$$

where $\varepsilon_{t,i}$ is the initial steel strain due to preloading at the bottom surface of the tension flange.

4.3.5 Moment–curvature responses

The moment-curvature response of FRP-plated steel/composite sections can be easily obtained from a section analysis based on the same principle as described above. To obtain a moment-curvature response in a step-by-step process, a series of values for the strain at the top surface of the section can be chosen for analysis. For a given strain value at the top surface of the beam, a neutral axis depth, x, can be found by solving the equilibrium equation: (1) Equation (4.4) for an FRP-plated steel section and (2) Equation (4.20) for an FRP-plated steel–concrete composite section.

For a given compressive strain $\varepsilon_{c,x}$ at the top surface of the beam and a given neutral axis depth x, the curvature of the section is given by

$$
\phi_x = \frac{\varepsilon_{c,x}}{x}
\tag{4.31}
$$

The moment resisted by the section is then given, for an FRP-plated steel beam, by

$$
M_{x,d} = \int_0^h \sigma_{s,i} b_i \left(\frac{h}{2} - h_{s,i}\right) dh_{s,i} + \sigma_{frp} b_{frp} t_{frp} \left(\frac{h}{2} + t_a + \frac{t_{frp}}{2}\right)
\tag{4.32}
$$

where the steel stress $\sigma_{s,i}$ at depth $h_{s,i}$ can be obtained from Eq. (4.5) or (4.6) using the strain $\varepsilon_{s,i}$ at depth $h_{s,i}$ obtained from Eqs. (4.1) and (4.3). σ_{frp} is the stress in the FRP laminate, which can be obtained from Eq. (4.10), where the FRP strain ε_{frp} can be obtained from Eq. (4.1).

For an FRP-plated steel–concrete composite section, the moment resisted by the section is

$$
M_{x,d} = k_1 \frac{f_{cu}}{\gamma_c} b_e x \left(\frac{h}{2} - k_2 x\right) + \int_{h-h_c}^h \sigma_{s,i} b_i \left(\frac{h}{2} - h_{s,i}\right) dh_{s,i}
$$
$$
+ \sigma_{frp} b_{frp} t_{frp} \left(\frac{h}{2} + t_a + \frac{t_{frp}}{2}\right)
\tag{4.33}
$$

where k_1 and k_2 can be obtained from (1) Eqs. (4.19) and (4.25), respectively, if $x \le h_c$, or (2) Eqs. (4.28) and (4.29), respectively, if $x > h_c$. The FRP strain and the steel strain, and in the case of $x > h_c$, the compressive strain at the bottom surface of the concrete slab, can be obtained from Eqs. (4.13), (4.14), and (4.27), respectively. The steel stress $\sigma_{s,i}$ can be obtained from Eq. (4.5) or (4.6), and the FRP stress σ_{frp} from Equation (4.10).

By steadily increasing the strain value at the top compression fibre and determining the moment and curvature for each strain increment, the moment-curvature response of a given section can be predicted. For beams failing by FRP tensile rupture, it makes more sense to use increments of the FRP strain. This can be easily achieved by determining the neutral axis depth for a given strain of the FRP laminate, and then determining the strain at the top compression fibre. The determination of the moment-curvature response then follows the same procedure as described above.

4.4 LATERAL BUCKLING

The flexural strengthening of a steel beam by a tension face FRP laminate is normally undertaken when in-plane bending rather than lateral buckling is the critical failure mode. This is because the benefit of a tension face FRP laminate is much more limited for the latter mode than for the former mode. This difference in the effect of a tension face FRP laminate also means that the critical failure mode of a steel beam may change from in-plane bending to lateral buckling. It is therefore important to consider the lateral buckling mode of an FRP-plated steel beam unless the lateral buckling mode is prevented by adequate lateral support.

There has been only very limited work on the lateral buckling of FRP-plated steel beams. A closed-form solution for the elastic lateral buckling load of FRP-plated steel beams has been developed by Zhang and Teng (2007). The effectiveness of different strengthening schemes for enhancing the lateral buckling resistance has also been explored by Kabir and Seif (2010). To the best of the authors' knowledge, no other study has been published on the topic.

As the existing research is far from adequate for the development of a simple method for evaluating the lateral buckling resistance of FRP-plated steel beams, designers are advised to resort to nonlinear finite element (FE) analysis. In such an analysis, both the steel beam and the FRP laminate can be represented using shell elements. The FRP laminate can be treated as an orthotropic thin plate, and assumed to be perfectly bonded to the steel beam, as debonding failure can be considered separately in design. The ultimate load of an FRP-plated steel beam failing by lateral buckling can be closely predicted by considering both geometric and material nonlinearities, as well as geometric imperfections and residual stresses. Such analyses for unplated steel beams have been done by many researchers (e.g., Pi and Trahair 1994, 1995). Designers should seek specialist advice when necessary in conducting such FE analyses. It should be noted that such a shell element-based model can also predict any local buckling failure, but for reliable predictions, appropriate geometric imperfections for the plate elements need to be included in the model.

4.5 DEBONDING FAILURES

4.5.1 General

As mentioned earlier, the two common debonding failure modes in FRP-plated steel or steel–concrete composite beams are end debonding and intermediate debonding. End debonding occurs due to high localised interfacial shear and normal (i.e., peeling) stresses at the laminate end, whereas intermediate debonding occurs in regions where the FRP laminate is highly stressed (often at a location of defect or yielding), mainly due to high interfacial shear stresses. Among these two debonding failure modes, end debonding has received more research attention, but much less is known about intermediate debonding failure.

In developing methods to predict debonding failures, it is important to note that such debonding can occur at the bimaterial interfaces between steel and adhesive or adhesive and FRP, or by failure within the adhesive layer with the bimaterial interfaces remaining intact. The former is commonly referred to as adhesion failure, while the latter is commonly referred to as cohesion failure. To maximise the benefit of the strengthening system and to facilitate the development of design methods, it is necessary to ensure that cohesion failure governs the debonding failure process (Teng et al. 2012). Such cohesion failure can be achieved through proper surface preparation (Fernando et al. 2013). In all discussions of debonding failure strength in this chapter, it is assumed that cohesion failure governs the debonding failure process.

4.5.2 Interfacial stresses in elastic FRP-plated beams

As end debonding in FRP-plated beams depends strongly on interfacial stress concentrations at plate ends, many studies have been published on the interfacial stress analysis of elastic plated beams in the context of strengthening beams of various materials. Both analytical (Roberts 1989, Taljsten 1997, Saadatmanesh and Malek 1998, Rabinovitch and Frostig 2000, Smith and Teng 2001, Shen et al. 2001, Colombi and Poggi 2006, Stratford and Cadei 2006, Yang and Ye 2010) and numerical (Teng et al. 2002b, Colombi and Poggi 2006, Linghoff and Al-Emrani 2010,

Zhang and Teng 2010a) investigations have been undertaken on this topic. These studies have been based on different simplifying assumptions corresponding to different levels of approximation as explained by Zhang and Teng (2010a). All the studies mentioned above are for linear-elastic beams where all components, including the adhesive layer, are assumed to behave in a linear-elastic manner. Any deviation from linear elasticity will greatly complicate the derivation of a closed-form analytical solution.

For exploitation in design, simple approximate closed-form solutions such as that developed by Smith and Teng (2001) may be used owing to their simplicity, compared to the more complicated higher-order solutions (e.g., Shen et al. 2001). These approximate solutions are based on the assumption that the interfacial stresses are constant across the adhesive layer thickness, but still provide a close representation of the overall interaction between the original beam and the bonded plate. Such approximate solutions have been exploited in attempts to predict end debonding by assuming that end debonding occurs when the maximum interfacial stresses found from such solutions reach their corresponding limiting stresses (i.e., material strengths) (Lenwari et al. 2006, Schnerch et al. 2007). This approach is conservative and may substantially underestimate the end debonding load (De Lorenzis et al. 2013).

The existing simple approximate solutions for interfacial stresses were generally developed for one or more specific loading conditions, which makes it difficult for them to be used in practical design where a variety of loading conditions are possible. More recently, Zhang and Teng (2010b) developed a simple general analytical solution for interfacial stresses in plated beams that is applicable to all loading conditions. In this solution, both the axial and bending deformations are included for both the original beam and the strengthening plate. Shear deformations of both components are neglected, as their effect is small. Based on this solution, the interfacial shear stress in a plated I-section beam with a constant cross section is given by

$$\tau(x) = \frac{s_1}{b_{frp}}\left[s_2 V(x) + \frac{dN(x)}{dx}\right] + \frac{s_1(s_2 M_0 + N_0)K_s}{E_{frp}A_{frp}\lambda}e^{-\lambda x} \qquad (4.34)$$

The subscript 0 is used to refer to the plate end section, i.e., $x = 0$ (Figure 4.13). M_0 and N_0 are the bending moment and the axial force acting on the plate end section, $V(x)$ and $N(x)$ are the shear force and the axial force acting on the section under consideration, and s_1 and s_2 are given by

$$s_1 = \frac{E_{frp}A_{frp}}{E_{frp}A_{frp} + E_1 A_1} \qquad (4.35)$$

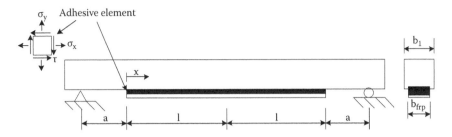

Figure 4.13 Simply supported plated beam and stresses acting on an adhesive element (interface) adjacent to a laminate end.

$$s_2 = \frac{(y_1 + y_2 + t_a)E_1A_1}{(EI)_{comp}} \tag{4.36}$$

$$(EI)_{comp} = E_1I_1 + E_{frp}I_{frp} + \frac{E_1A_1E_{frp}A_{frp}(y_1 + y_2 + t_a)^2}{E_1A_1 + E_{frp}A_{frp}} \tag{4.37}$$

where y_1 and y_2 are the distances from the bottom surface of the bottom flange and the top surface of the FRP plate to their respective neutral axes, and K_s is the shear stiffness of the adhesive layer, given by

$$K_s = \frac{G_a}{t_a} \tag{4.38}$$

and λ is given by

$$\lambda = \sqrt{\frac{K_s}{E_{frp}t_{frp}}} \tag{4.39}$$

The solution for the interfacial normal stress is given by

$$\sigma_n(x) = -K_n e^{-\beta x}\left[D_1 \cos(\beta x) + D_2 \sin(\beta x)\right] + \frac{d\tau(x)}{dx}y_2 \tag{4.40}$$

in which

$$K_n = \frac{E_a}{t_a} \tag{4.41}$$

$$\beta^4 = \frac{K_n b_2}{4E_{frp}I_{frp}} \tag{4.42}$$

$$D_1 = -\frac{V_{20} + \beta M_{20}}{2E_{frp}I_{frp}\beta^3} \qquad (4.43)$$

$$D_2 = \frac{M_{20}}{2E_{frp}I_{frp}\beta^2} \qquad (4.44)$$

$$M_{20} = M_0\left(\frac{E_{frp}I_{frp}}{E_1I_1 + E_{frp}I_{frp}}\right) \qquad (4.45)$$

$$V_{20} = \frac{\tau_0 A_{frp}}{2} \qquad (4.46)$$

For the three loading conditions shown in Figures 4.14(a) to (c), the expressions for the shear force and the bending moment are listed in Table 4.1, where $L = l + a$. The axial force, $N(x)$, is zero for all three loading conditions.

The typical interfacial shear and normal stress distributions predicted by the above equations are shown in Figure 4.15. Both the maximum normal stress and the maximum shear stress are predicted to occur at the laminate end where debonding failure is expected.

4.5.3 Cohesive zone modelling of debonding failure

As mentioned above, the direct use of the peak values of interfacial stresses to predict end debonding is likely to underestimate significantly the end debonding load. A more accurate prediction of end debonding can be made using the so-called cohesive zone (CZ) model for the interface subjected to

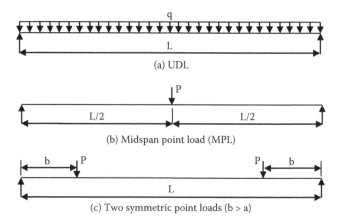

(a) UDL

(b) Midspan point load (MPL)

(c) Two symmetric point loads (b > a)

Figure 4.14 Three common loading conditions (L = l + a).

Table 4.1 Expressions for shear forces and bending moments for different loading conditions

Load condition	Shear force, $V(x)$	Moment, $M(x)$
UDL (Figure 4.14a)	$q\left(\dfrac{L}{2}-x-a\right)\ 0\le x<\dfrac{L}{2}-a$	$\dfrac{qL}{2}(x+a)-\dfrac{q}{2}(x+a)^2\ 0\le x<\dfrac{L}{2}-a$
Point load (Figure 4.14b)	$\dfrac{P}{2}\ \ 0\le x<\dfrac{L}{2}-a$	$\dfrac{P}{2}(a+x)\ \ 0\le x<\dfrac{L}{2}-a$
Two point loads (Figure 4.14c)	$\begin{cases} P & 0\le x<b-a \\ 0 & b-a\le x<\dfrac{L}{2} \end{cases}$	$\begin{cases} P(x+a) & 0\le x<b-a \\ Pb & b-a\le x<\dfrac{L}{2} \end{cases}$

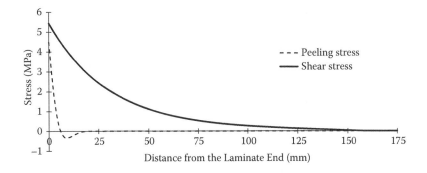

Figure 4.15 Typical interfacial shear and normal stress distributions.

combined normal stresses and shear stresses. The same approach can also be used to predict intermediate debonding arising from a local damage in an FRP-plated steel/composite beam.

Prediction of debonding failures in FRP-plated steel beams using the CZ model requires a constitutive law for the FRP-to-steel interface, which has the following characteristics: (1) it accurately predicts the nonlinear and damage behaviour of the interface subjected to either pure mode I loading or pure mode II loading, and (2) it appropriately accounts for the effect of interaction between mode I loading and mode II loading on damage propagation along the interface. For (1), an accurate bond–slip model (e.g., such as those presented in Fernando 2010) needs to be employed to predict the full-range interfacial behaviour under pure mode II loading, where the area under the bond–slip curve represents the mode II fracture energy. Similarly, the full-range interfacial behaviour under pure mode I loading can be depicted by a bond-separation model (de Moura and Chousal 2006, Yuan and Xu 2008), where the area under the bond-separation curve represents the fracture energy under mode I loading. For (2), the so-called mixed-mode cohesive law needs to be employed. Among the existing techniques available for modelling mixed-mode debonding, a coupled CZ model appears to be the most suitable, as it possesses both of the two required characteristics.

CZ models have been used in simulating the fracture of ductile and brittle solids (De Lorenzis and Zavarise 2009), the delamination of composite materials (Sorensen 2002, Sorensen and Jacobsen 2003, Li et al. 2005, 2006), and the behaviour of adhesively bonded joints (Goncalves et al. 2003, Liljedahl et al. 2006). Bocciarelli et al. (2007) presented the first FE model for CFRP-to-steel bonded joints with a CZ model and showed good predictions for double-lap shear tests, but the CZ model used by them considered interfacial behaviour under pure mode II loading only. Fernando (2010) proposed a coupled CZ model for CFRP-to-steel interfaces, which includes all three important components: a bond–slip model for mode II loading, a bond-separation model for mode I loading, and a coupled law for mixed-mode behaviour. The mixed-mode CZ model proposed by Fernando (2010) was shown to be successful in predicting both intermediate debonding and end debonding of FRP-plated steel beams under a mid-span point load (MPL).

4.5.4 End debonding

4.5.4.1 General

Fernando (2010) demonstrated the effectiveness of using a mixed-mode cohesive law to predict the end debonding strength of CFRP-plated steel beams. De Lorenzis et al. (2013) proposed a simplified analytical model based on the same mixed-mode cohesive law to predict the end debonding strength of FRP-plated elastic steel beams. In the remainder of this subsection, both approaches are briefly described. The FE model of Fernando (2010) is superior to the analytical model of De Lorenzis et al. (2013) in terms of general applicability and accuracy, while the analytical model of De Lorenzis et al. (2013) is computationally simple and can be implemented using readily available software such as Microsoft Excel.

4.5.4.2 FE modelling

Fernando (2010) presented a detailed FE study on the use of a coupled CZ model to predict interfacial debonding in CFRP-plated steel beams. The FE model was verified using the experimental data from Deng and Lee (2007a). Figure 4.16, taken from Fernando (2010), shows comparisons between test results and FE predictions based on the CZ model for two

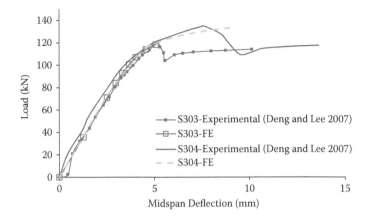

Figure 4.16 FE versus test results for end debonding. (From Fernando, D., Bond Behaviour and Debonding Failures in CFRP-Strengthened Steel Members, PhD thesis, Hong Kong Polytechnic University, Kowloon, Hong Kong, 2010.)

beams (S303 and S304 of Deng and Lee 2007a). It is clear that the proposed FE model is capable of accurately predicting the end debonding strength. Therefore, this CZ model given in Fernando (2010) may be employed to predict end debonding in design where appropriate.

4.5.4.3 Analytical modelling

Using the coupled CZ model described in Section 4.5.3, De Lorenzis et al. (2013) proposed an analytical model to predict the end debonding strength of FRP-plated steel beams under an MPL. Due to the complexity of modelling, the developed analytical model is only strictly applicable to FRP-plated steel beams where end debonding occurs before the yielding of the steel beam. In practice, it is highly undesirable to allow end debonding to occur before the steel beam has yielded at all, as this implies that the material strength of FRP is far from being fully utilised, and the contribution of FRP to the flexural strength of the beam is small. The analytical solution of De Lorenzis et al. (2013) can, however, provide a relatively simple and approximate tool for predicting the end debonding strength for beams where yielding has not spread to the vicinities of the plate ends. Although the analytical model of De Lorenzis et al. (2013) is limited to the single loading condition of MPL, their approach does have the potential for extension to other loading conditions. It is worthwhile to undertake further research along this direction to develop a more general design model for end debonding strengths. Predictions from De Lorenzis et al.'s (2013) model have been shown by the authors to agree well with FE predictions from the CZ model.

4.5.4.4 Suppression through detailing

Even if the end debonding strength can be properly predicted, end debonding may occur so early that the material strength of FRP is far from fully utilised, which leads to inefficient use of the expensive FRP material. The better approach in practical applications is to suppress end debonding by (1) placing the laminate end in a low-moment region and (2) providing mechanical anchorage such as FRP wraps at the laminate end. Other methods that have been suggested to reduce the risk of end debonding include (3) using an adhesive with a large fracture energy (often a soft adhesive) at the laminate end (Fitton and Broughton 2005), (4) reducing end stress concentrations by providing a spew fillet of excess adhesive, and (5) tapering the edge of the bonded FRP laminate (Hart-Smith 1981, Haghani et al. 2010) (Figure 4.17(a)). All these methods can be effective to different extents in delaying or suppressing end debonding failure, but methods 4 and 5 are believed to have limited benefits, as end debonding is not stress controlled. In addition, the use of a combination of some of these methods, such as providing a spew fillet of excess adhesive and reverse tapering of the laminate thickness near the laminate end, has been suggested (Schnerch et al. 2007).

Where possible, the provision of clamps (Figure 4.17(b)) or other types of mechanical anchors to avoid end failure due to high peeling stresses is advisable (Sen et al. 2001). It is recommended herein that FRP wraps (Figure 4.17(c) and (d)) should be provided to suppress end debonding unless access to the sides of the beam is blocked. An FRP end wrap can be so proportioned that it is able to resist the peeling stresses at the end, and these peeling stresses can be estimated using the simple general analytical solution of Zhang and Teng (2010b) presented earlier. The total peeling force to be resisted by an FRP end wrap (a U-jacket in the case of an RHS beam) can be found by integrating the normal stresses within the vicinity of the laminate end (i.e., the region where tensile normal stresses exist) under the ultimate load, i.e.,

$$P_n = b_{frp} \int_0^{x|_{\sigma_n(x)=0}} \sigma_n(x)\,dx \qquad (4.47)$$

where P_n is the total peeling force to be resisted by an FRP end wrap, and $x|_{\sigma_n(x)=0}$ is the distance from the plate end to the first point where the interfacial normal stress becomes zero.

The use of the elastic interfacial stress analysis can be justified by noting that the laminate end regions of the beam are likely to be still in an elastic state under the ultimate load, and that the elastic stress analysis is expected to lead to a conservative design.

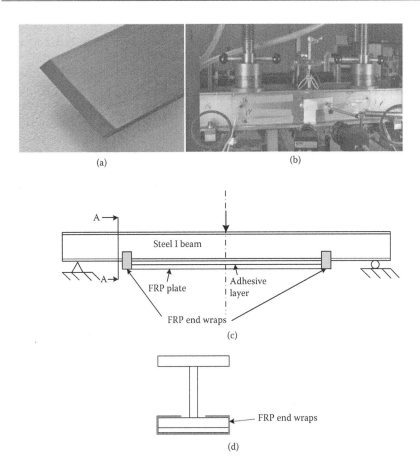

Figure 4.17 End details for suppression of end debonding. (a) End tapering of CFRP laminate. (Reprinted from Schnerch, D., and Rizkalla, S., *Journal of Bridge Engineering*, ASCE, 13(2), 192–201, 2008.) (b) Mechanical anchors. (c) FRP end wraps. (d) Section A-A of FRP plated I-section beam (in Figure 4.17c).

4.5.5 Intermediate debonding

Intermediate debonding generally initiates at a local damage such as a crack (Kim and Garrett 2011, Kim and Kent 2012) or localised yielding (Sallam et al. 2006), where the FRP laminate is highly stressed; it then propagates toward a laminate end. Although both end debonding and intermediate debonding are brittle failure modes, the latter, involving a more gradual process of debonding, is generally less brittle than the former (Fernando 2010). Compared with end debonding, much less research is available on intermediate debonding in FRP-plated steel/composite beams (Fernando 2010).

Intermediate debonding in FRP-plated steel/composite beams is similar in nature to intermediate crack (IC) debonding in FRP-plated concrete beams (Teng et al. 2003): both initiate from a location where the FRP is highly stressed, and both are dominated by interfacial shear stresses. Therefore, it can be expected that the intermediate debonding strength depends strongly on the interfacial shear fracture energy obtained from shear tests on bonded joint tests (Fernando 2010). By assuming that intermediate debonding occurs at the same strain as found from bonded joint shear tests, the design value of the FRP strain at intermediate debonding can be expressed as

$$\varepsilon_{frp,I,d} = \frac{F_{I,d}}{\gamma_{debond} b_{frp} E_{frp} t_{frp}} = \frac{1}{\gamma_{debond}} \sqrt{\frac{2G_f}{E_{frp} t_{frp}}} \quad \text{for } L \geq L_e \tag{4.48}$$

where λ_{debond} is the partial safety factor for debonding failure, b_{frp}, E_{frp}, and t_{frp} are the width, elastic modulus, and thickness of the FRP laminate, respectively, and for both linear and nonlinear adhesives, the interfacial fracture energy, G_f, is given by (Fernando 2010)

$$G_f = 628 t_a^{0.5} R^2 \tag{4.49}$$

where t_a is the adhesive layer thickness in mm and R is the tensile strain energy of the adhesive, which is equal to the area under the uniaxial tensile stress (in MPa)-strain curve.

The direct application of Eq. (4.48) is expected to provide close predictions only for intermediate debonding due to a single localised damage in the steel substrate. Where intermediate debonding may be initiated at multiple locations, direct application of this equation should lead to conservative predictions due to interactions between adjacent locations (Teng et al. 2006). The design moment capacity of an FRP-plated beam at intermediate debonding can be calculated using the appropriate equations from Section 4.3, by replacing $\varepsilon_{frpl,d}$ with $\varepsilon_{frp,I,d}$ obtained from Eq. (4.48).

In using Eq. (4.48), the bond length needs to be not shorter than a minimum bond length (referred to as the effective bond length). For interfaces with a linear adhesive, the bond–slip curve is assumed to be bilinear; this effective bond length is given by (Fernando 2010)

$$L_e = a_b + \frac{1}{2\lambda_1} \ln\left(\frac{\lambda_1 + \lambda_2 \tan(\lambda_2 a_b)}{\lambda_1 - \lambda_2 \tan(\lambda_2 a_b)}\right) \tag{4.50a}$$

with

$$a_b = \frac{1}{\lambda_2}\arcsin\left[0.97\sqrt{\frac{\delta_f - \delta_1}{\delta_f}}\right] \tag{4.50b}$$

$$\lambda_1^2 = \frac{2G_f}{\tau_{max}\delta_1}\lambda^2 \tag{4.50c}$$

$$\lambda_2^2 = \frac{2G_f}{(\delta_f - \delta_1)\tau_{max}}\lambda^2 \tag{4.50d}$$

$$\lambda^2 = \frac{\tau_{max}^2}{2G_f}\left(\frac{1}{E_{frp}t_{frp}} + \frac{b_{frp}}{E_sA_s}\right) \tag{4.50e}$$

where δ_1 and δ_f are respectively the slip at peak shear stress and the maximum slip of the bond–slip model (see Fernando (2010) for details of the bilinear bond–slip model).

For interfaces with a nonlinear adhesive, the bond–slip curve is assumed to be trapezoidal, and the effective bond length is given by (Fernando 2010)

$$L_e = a_d + b_e + \frac{1}{\lambda_1}\ln\frac{1+C}{1-C} \tag{4.51a}$$

with

$$C = \frac{\lambda_3}{\lambda_1\delta_1}(\delta_f - \delta_2)\cot(\lambda_3 b_e) - \lambda_1 a_d \tag{4.51b}$$

$$b_e = \frac{1}{\lambda_3}\arcsin\left[\frac{\lambda_3\lambda}{0.97\delta_1\lambda_1^2}(\delta_f - \delta_2)\right] \tag{4.51c}$$

$$a_d = \frac{1}{\lambda_1}\left(\sqrt{\left(2\frac{\delta_2}{\delta_1} - 1\right)} - 1\right) \tag{4.51d}$$

$$\lambda_3^2 = \lambda^2\frac{2G_f}{(\delta_f - \delta_2)\tau_{max}} \tag{4.51e}$$

$$\lambda_1^2 = \frac{2G_f}{\tau_{max}\delta_1}\lambda^2 \tag{4.51f}$$

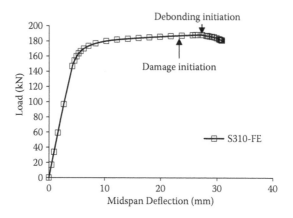

Figure 4.18 Load-deflection curve from an FE model for beam S310-1-212-P.

$$\lambda^2 = \frac{\tau_{max}^2}{2G_f}\left(\frac{1}{E_{frp}t_{frp}} + \frac{b_{frp}}{E_sA_s}\right) \tag{4.51g}$$

δ_1, δ_2, and δ_f are the slip values at the end of the linear ascending branch, at the beginning of the linear descending branch, and at the end of the linear descending branch, respectively; these values can be obtained as described in Fernando (2010).

Fernando (2010) showed that with the use of a bond–slip model for mode II interfacial fracture, both the process and the ultimate load of intermediate debonding can be closely predicted using the FE method. Therefore, in principle, the numerical approach explained in Section 4.5.4.2 is also applicable to the prediction of intermediate debonding. The load-deflection curve of an FRP-plated steel beam (beam S310-1-212P in Fernando 2010: 127x76UB13 steel beam of 1000 mm in length, with a 76 × 3 mm CFRP laminate bonded to the tension flange and a 6 mm thick steel plate welded to the compression flange) from such an FE model is given in Figure 4.18.

4.5.6 Local buckling

4.5.6.1 Design against flange and web buckling

As briefly discussed in Section 4.2.3, even in compact steel/composite beams strengthened with FRP laminates bonded to the tension flange, it is possible that post-yielding (i.e., inelastic) compression flange buckling or web buckling occurs due to increased compressive straining. When applying the design equations in Section 4.3, it is necessary to ascertain whether inelastic local buckling occurs before other modes of failure. As inelastic local buckling can lead to a significant reduction in the beam section capacity, it constitutes an ultimate limit state.

Most of the existing design codes do not contain a method for checking the post-yielding buckling of plate elements in steel/composite beams, and no study has been published on the problem. An approximate method is proposed herein for beams with a steel I-section, based on the work of Aoki and Kubo (1997), where a rotational capacity equation is given to design against the compression flange buckling and web buckling. For more accurate predictions of such local buckling in FRP-plated steel/ composite beams, designers are advised to conduct nonlinear finite element (FE) analysis based on modelling techniques similar to those explained in Section 4.4.

According to Aoki and Kubo (1997), the maximum rotational capacity of an I-section steel beam is given by

$$\phi_{lb} = \frac{0.15}{\lambda_s - 0.36} \leq 12 \tag{4.52a}$$

$$\lambda_s = \sqrt{\lambda_{pf}\lambda_{pw}} \tag{4.52b}$$

where λ_{pf} and λ_{pw} are the width-to-thickness ratio parameters of flange and web, respectively:

$$\lambda_p = \frac{b}{t} \sqrt{\frac{12(1-v^2)}{\pi^2 k}} \sqrt{\frac{\sigma_{s,y}}{E_s}} \tag{4.53}$$

In the above equation, b is the width and t the thickness of the plate element (flange or web), and k is the buckling coefficient, which is taken as 0.43 for flanges and 2.39 for webs.

For an FRP-plated steel section, λ_{pw} can be obtained using an equivalent section as illustrated in Figure 4.19.

The design against web shear buckling can be undertaken using appropriate methods given in existing design codes, and is thus not discussed in detail here. Web shear buckling can be deemed to be noncritical if the following condition is satisfied (EN 1993 2005):

$$\frac{(h - 2t_f)}{t_w} < 72\frac{\varepsilon}{\eta} \tag{4.54}$$

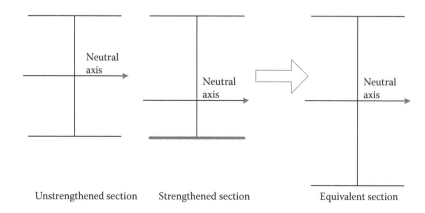

Figure 4.19 Equivalent section for calculating plate slenderness.

where $\varepsilon = \sqrt{\dfrac{235}{\sigma_{s,y}}}$ and η may be conservatively taken to be 1 (EN 1993 2005).

4.5.6.2 Additional strengthening against local buckling

In order to obtain the maximum benefit of FRP strengthening of the tension flange, it may be necessary to provide additional strengthening of the compression flange and the web against local buckling.

Attempts have been made to use FRP composites to strengthen steel plates against local buckling (Zhao et al. 2006, Fernando et al. 2009). These include bonding of FRP laminates (Zhao et al. 2006, Fernando et al. 2009) as well as FRP sections (Okeil et al. 2009). Zhao et al. (2006) showed that bonded CFRP laminates can strengthen the webs of RHS beams against web crippling effectively. The effectiveness of such a strengthening system depends significantly on the properties of the adhesive used; adhesives with a larger ultimate tensile strain/fracture energy lead to greater strength increases (Fernando et al. 2009, Fernando 2010).

Simple but generally applicable methods for predicting the local buckling failure of such FRP-strengthened plates have not yet been developed, but Fernando (2010) showed that an FE model based on the mixed-mode CZ approach, as described in Section 4.5.4.2, can yield accurate results for CFRP-strengthened RHS tubes under an end bearing load. Therefore, in principle this FE approach can be applied to predict the load carrying capacity of plate elements strengthened using FRP laminates against local buckling. Fernando (2010) demonstrated the capability of this numerical approach for linear adhesives, but similar work has not been undertaken for nonlinear adhesives.

4.6 OTHER ISSUES

4.6.1 Strengthening of beams without access to the tension flange surface

The discussions presented in this chapter so far are focussed on the flexural strengthening of steel/composite beams whose tension flange can be accessed for the bonding of FRP laminates. However, there are situations where the outer surface of the tension flange is not accessible (e.g., negative moment regions). In such situations, removing the covering material to expose the tension flange is generally neither feasible nor desirable. In such cases, the FRP laminate can be bonded to the inner side of the tension flange. The moment capacity of such an FRP-strengthened beam can be found from a section analysis similar to that described in Section 4.3, with due account taken of the slightly different position of the FRP laminate.

4.6.2 Rapid strengthening methods

In the strengthening of steel/composite bridges, the speed of strengthening operation is of great importance in minimising/avoiding traffic closure. Hollaway et al. (2006) and Zhang et al. (2006) investigated the rapid strengthening of steel bridges using prepregs and a film adhesive. Using this new method, a bridge may be strengthened in as little as 4 hours. They also examined the effect of traffic-induced vibration during the curing of the FRP system on the performance of the strengthened structure. The effectiveness and reliability of this rapid strengthening method for steel structures were demonstrated by Hollaway et al. (2006).

4.6.3 Fatigue strengthening

The bonding of FRP laminates to improve the fatigue life of steel structures is an attractive strengthening method (Bassetti et al. 2000a, 2000b, Colombi et al. 2003a, Dawood et al. 2007, Liu et al. 2009a, 2009b, Jiao et al. 2012). Studies on fatigue strengthening using FRP laminates have been carried out on steel beams (Tavakkolizadeh and Saadatmanesh 2003b, Deng and Lee 2007b, Shaat and Fam 2008, Eglimez and Yormaz 2011, Ghafoori et al. 2012), steel plates (Colombi et al. 2003b, Nakamura et al. 2008, Bocciarelli et al. 2009, Liu et al. 2009a, 2009b, Colombi and Giulia 2012, Wu et al. 2012), steel rods (Jones and Civjan 2003), and steel connections (Nakamura et al. 2009, Chen et al. 2012, Xiao and Zhao 2012).

Similar to the behaviour of FRP-to-steel bonded joints under static loading, Liu et al. (2009a, 2009b) found that the fatigue life of FRP-strengthened steel plates initially increased with the bond length until the effective bond length under cyclic loading $L_{e,f}$ (there is not enough research on the difference between this effective bond length and the effective bond length under static loading) was reached, after which any further increase in the bond length did not further increase the fatigue life. In the strengthening of steel members (e.g., plates, beams, and rods), a bond length longer than $L_{e,f}$ is strongly advised.

Stress intensity factors (SIFs) are commonly used to describe the stress state at a crack tip due to applied loading or residual stresses (Colombi et al. 2003b, Taljsten et al. 2009). Fatigue strengthening of steel structures generally aims to reduce the SIF at a (potential) crack tip, and thus increase their post-crack fatigue life. In addition to experimental work, a number of analytical studies (Colombi et al. 2003a, Liu et al. 2009a, Tsouvalis et al. 2009) have been conducted on the prediction of SIFs at crack tips in FRP-strengthened steel structures. Such analysis is necessary and useful in the design of FRP systems for the fatigue strengthening of steel structures. As may be expected, the use of a stiffer FRP laminate (i.e., with a greater thickness or a higher elastic modulus) or a stiffer adhesive (i.e., with a higher elastic modulus) leads to greater reductions in the SIF (Liu et al. 2009a, 2009b, Nakamura et al. 2009).

Under fatigue loading, debonding of FRP laminates initiating at cracks in a strengthened member can be expected, and such debonding can significantly increase the SIF, and thus compromise the benefit of fatigue strengthening (Colombi et al. 2003a, Shaat and Fam 2008). In the existing literature, debonding of FRP laminates from the steel substrate on the effectiveness of fatigue strengthening has received very limited attention; the only existing work has been concerned with the effect of a debonded zone of prescribed shape and size, which is a function of the substrate crack width, on the SIF (Megueni et al. 2004, Ouinas et al. 2007). More research is certainly needed to gain a better understanding of the cyclic behaviour of CFRP-to-steel bonded interfaces, as well as the interaction between intermediate debonding and fatigue crack growth in steel beams, so that the detrimental effect of debonding on the fatigue life of CFRP-strengthened steel/composite beams can be predicted.

Prestressing the FRP laminate can significantly enhance the effectiveness of fatigue strengthening. By pretensioning the FRP laminate, compressive stresses are induced in the steel substrate to achieve crack closure, resulting in improved fatigue performance. The effect of the pretensioning level on the fatigue crack growth rate has been studied both experimentally and numerically (Colombi et al. 2003a, 2003b, Taljsten et al. 2009). By evaluating the SIF at the crack tip of the strengthened system, the pretensioning force needed to stop the growth of a fatigue crack can be predicted (Taljsten et al. 2009). The level of pretensioning that can be imposed on an FRP strengthening system depends on the static and fatigue strengths of the bonded joint, where a good understanding of the behaviour of bonded interfaces is again required.

4.7 DESIGN RECOMMENDATION

4.7.1 General

Based on the information presented in the preceding sections of this chapter, a comprehensive design procedure can be established for the static strengthening of steel/composite beams using FRP laminates bonded to the tension flange. Such a design procedure is set out below. Existing research is inadequate for the establishment of reasonable design recommendations for other static strengthening schemes and for fatigue strengthening.

4.7.2 Critical sections and end anchorage

In determining the ultimate load of an FRP-plated steel/composite beam, all possible failure modes, including debonding and buckling, need to be considered. Debonding failures can be designed against using section analysis by equating the FRP limiting strain to the debonding strain, and this section capacity check needs to be conducted for two critical sections: (1) the maximum moment section and (2) the section just outside the effectively strengthened region, provided that the moment capacity of the original beam does not vary along the beam length. The effectively strengthened region is the region within which the section capacity is inadequate without the FRP contribution when the plated beam is under its ultimate load. If the moment capacity of the original beam varies along the length, then in principle, section capacity checks are necessary at all sections with a capacity change.

To minimise the risk of end debonding, the FRP laminate should be terminated as close to the supports as possible. A small distance between the laminate end and the adjacent support is, however, always expected, as the underside of the beam near the location of the idealised support is generally inaccessible due to the presence of the physical support for the beam. The distance between the laminate end and a critical section (i.e., the anchorage length) needs to be sufficiently long to achieve a reliable transfer of the tensile force in the FRP laminate to the steel beam. It is recommended herein that this anchorage length should at least be equal to twice the effective bond length evaluated by either Eq. (4.50a) for linear adhesives or Eq. (4.51a) for nonlinear adhesives. This anchorage length requirement is to ensure that if intermediate debonding induced by a local defect or localised yielding occurs at a critical section, the tensile force that can be sustained by the FRP laminate is maximised.

In addition to the minimum anchorage length specified above, end anchors such as FRP end wraps (Figure 4.20) are recommended to suppress end debonding. Such wraps should be proportioned to resist the total peeling force found by integrating the peeling stresses near the laminate end (Eq. (4.47)) and by assuming that the tensile rupture of FRP wraps occurs

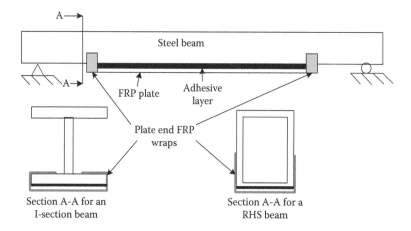

Figure 4.20 End wraps for an I-section steel beam and an RHS steel beam.

at one-third of their design tensile strength determined from coupon tensile tests. The latter is recommended to account for the substantial reduction in the tensile rupture strength of FRP wraps when they are formed into sharp corners. The interfacial normal stresses should be found using an appropriate numerical or analytical model. Zhang and Teng's (2010b) solution is recommended for use here due to its wide applicability and simplicity. The width of the end wrap is proposed to be the same as the FRP laminate bonded to the beam tension face, but this recommendation is purely based on the judgement of the authors.

The design moment resistance of the beam section outside the effectively strengthened region can be found using rules of an existing code of practice, such as EN 1993 (2005) for steel beams or EN 1994 (2002) for steel–concrete composite beams. To check the strength of the maximum moment section strengthened with a bonded FRP laminate, the following procedure is recommended.

4.7.3 Strength of the maximum moment section

The design value of moment capacity of an FRP-plated section, $M_{u,d}$, is given by

$$M_{u,d} = \min (M_{ib,d}, M_{ftb,d}) \qquad (4.55)$$

where $M_{ib,d}$ is the design value of section moment capacity due to in-plane failure, and $M_{ftb,d}$ is the design value of lateral (i.e., flexural–torsional) buckling moment capacity (the maximum moment in the beam when lateral buckling occurs; only applicable to FRP-plated steel beams).

4.7.3.1 Moment capacity at in-plane failure

To determine the in-plane design moment capacity, $M_{ib,d}$, of the FRP-plated section, the design value of FRP limiting strain, $\varepsilon_{frpl,d}$, should first be determined as follows:

$$\varepsilon_{frpl,d} = \min\left(\varepsilon_{frp,rup,d}, \varepsilon_{frp,I,d}, \varepsilon_{frp,E,d}\right) \tag{4.56}$$

where $\varepsilon_{frp,rup,d}$ is the design value of tensile rupture strain of FRP, $\varepsilon_{frp,I,d}$ is the design value of FRP strain at intermediate debonding, which is given by Eq. (4.48), and $\varepsilon_{frp,E,d}$ is the design value of FRP strain at end debonding, which can be evaluated in accordance with Section 4.5.4.2.

With the limiting strain of the FRP laminate determined, the in-plane moment capacity of the FRP-plated section can be found for (i) an FRP-plated steel section using Eq. (4.11) and (ii) an FRP-plated steel–concrete composite section using Equation (4.26), with the neutral axis position and the failure mode (i.e., FRP rupture or concrete crushing) properly considered.

4.7.3.2 Moment capacity at lateral buckling failure

The design moment capacity at lateral buckling failure of an FRP-plated steel beam, $M_{ftb,d}$, should be determined conservatively using an existing code of practice (e.g., EN 1993 2005) by assuming that the contribution of the FRP laminate can be ignored. More research is needed before a design method can be formulated for the accurate evaluation of the lateral buckling load of FRP-plated steel beams with the contribution of the FRP laminate properly accounted for. If the conservation evaluation indicates that lateral buckling is the critical failure mode, then the designer may need to resort to nonlinear FE analysis for an accurate evaluation.

4.7.3.3 Design against local buckling

When $M_{u,d}$ is determined, the following checks should be made:

1. Check for compression flange and web buckling.
   a. Determine the maximum compression fibre strain and the neutral axis depth when the section is subjected to its ultimate moment.
   b. Check for flange buckling using appropriate FE analysis or using the beam rotation capacity method as described in Section 4.5.6.1.
   c. If flange buckling is found to be critical, FRP strengthening of the flange needs to be specified and designed based on nonlinear FE analysis.
2. Check for web shear buckling.
   a. Determine the shear force and bending moment acting on the web when the section is subjected to its ultimate moment.
   b. Check for web shear buckling using any existing code of practice, such as EN 1993 (2005).

c. If web shear buckling is found to govern the strength of the beam, FRP strengthening of the web needs to be specified and designed based on nonlinear FE analysis.

4.8 DESIGN EXAMPLE

4.8.1 Geometric and material properties of the beam

To illustrate the application of the design procedure presented in the preceding section, a design example is presented in this section for a steel beam. The dimensions of the steel beam are given in Figure 4.21. The beam is simply supported with a span of 4000 mm. A normal-modulus pultruded CFRP plate of 1.4 mm in thickness, 166 mm in width, and 3900 mm in length is to be bonded to the tension flange of the beam. The beam is subjected to a point load at the mid-span. The material properties of steel and CFRP (in the fibre direction) and the corresponding partial safety factors suggested for use in this design example are given in Table 4.2. The beam is considered to be laterally restrained, and thus lateral buckling is ignored.

A nonlinear adhesive with the following properties is specified: elastic modulus = 1.75 GPa, tensile strength = 1.47 × 10 MPa, tensile strain energy = 1.48 × 10^{-1} MPa.mm/mm. The mode II behaviour of the adhesive layer can be described by a trapezoidal bond–slip curve with the following slip values: δ_1 = 1.80 × 10^{-2} mm, δ_2 = 8.00 × 10^{-1} mm, and δ_f = 1.28 mm.

4.8.2 In-plane moment capacity of plated section

Step 1: Determine the design limiting strain of the FRP plate:

$$\varepsilon_{frpl,d} = \min(\varepsilon_{frp,rup,d}, \varepsilon_{frp,I,d})$$

a. Design rupture strain of the FRP plate:

$$\varepsilon_{frp,rup,d} = \frac{2.80 \times 10^3}{1.25 \times 1.65 \times 10^5} = 1.36 \times 10^{-2}$$

b. Design value of FRP strain at intermediate debonding:

$$G_f = 6.28 \times 10^2 \times (1.00)^{0.50} \times 1.48 \times 10^{-1} = 1.37 \times 10 \text{ Nmm/mm}^2$$

$$\varepsilon_{frp,I,d} = \frac{1.00}{1.25} \times \sqrt{\frac{2.00 \times 1.37 \times 10}{1.65 \times 10^5 \times 1.40}} = 8.70 \times 10^{-3} \quad \text{for} \quad L \geq L_e$$

c. Check the anchorage length requirement.

The critical section of the beam is at the mid-span, so the anchorage length of the plate is:

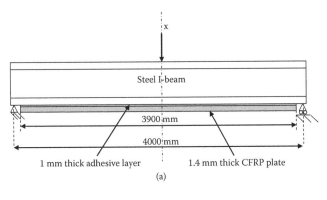

Steel I-beam

3900 mm

4000 mm

1 mm thick adhesive layer   1.4 mm thick CFRP plate

(a)

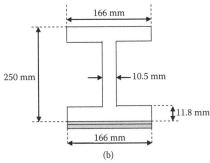

166 mm

250 mm   ←10.5 mm

↕11.8 mm

166 mm

(b)

Figure 4.21 Details of the example FRP-plated steel beam. (a) Beam loading configuration. (b) Cross section of CFRP-plated beam.

Table 4.2 Material properties of steel and CFRP and suggested values for partial safety factors

Material	Elastic modulus (GPa)	Strength (MPa)	Partial safety factor
Steel	2.00×10^2	2.75×10^2	$\gamma_s = 1.0$[a]
CFRP	1.65×10^2	2.80×10^3	$\gamma_{frp} = 1.25$
Bond			$\gamma_{debond} = 1.25$

[a] Taken as 1.0 according to EN 1993 (2005). The steel strength is taken as the steel yield strength of grade S275 steel as specified in EN 1993 (2005).

$$L_a = \frac{3.90 \times 10^3 \, mm}{2.00} = 1.95 \times 10^3 \, mm$$

$$\lambda = \sqrt{\frac{\left(9.00 \times 10^{-1} \times 1.47 \times 10\right)^2}{2.00 \times 1.37 \times 10} \times \left(\frac{1.00}{1.65 \times 10^5 \times 1.40} + \frac{1.66 \times 10^2}{2.00 \times 10^5 \times 6.29 \times 10^3}\right)}$$

$$= 5.36 \times 10^{-3} \, mm^{-1}$$

$$\lambda_1 = \sqrt{\frac{2.00 \times 1.37 \times 10}{(9.00 \times 10^{-1} \times 1.47 \times 10) \times 1.80 \times 10^{-2}}} \times 5.36 \times 10^{-3}$$

$$= 5.73 \times 10^{-2}\,\mathrm{mm}^{-1}$$

$$\lambda_3 = \sqrt{\frac{2.00 \times 1.37 \times 10}{(9.00 \times 10^{-1} \times 1.47 \times 10) \times (1.28 - 8.00 \times 10^{-1})}} \times 5.36 \times 10^{-3}$$

$$= 1.11 \times 10^{-2}\,\mathrm{mm}^{-1}$$

$$a_d = \frac{1}{5.73 \times 10^{-2}} \times \left(\sqrt{2 \times \frac{8.00 \times 10^{-2}}{1.80 \times 10^{-2}} - 1} - 1 \right) = 1.46 \times 10^2 \ \mathrm{mm}$$

$$b_e = \frac{1.00}{1.11 \times 10^{-2}} \arcsin\left[\frac{1.11 \times 10^{-2} \times 5.36 \times 10^{-3}}{9.70 \times 10^{-1} \times (5.73 \times 10^{-2})^2} \times (1.28 - 8.00 \times 10^{-1}) \right]$$

$$= 4.68 \times 10 \ \mathrm{mm}$$

$$C = \frac{5.73 \times 10^{-2}}{5.36 \times 10^{-3} \times 1.80 \times 10^{-2}} \times (1.28 - 8.00 \times 10^{-1})$$

$$\times \cot(5.73 \times 10^{-2} \times 4.68 \times 10) - 5.36 \times 10^{-3} \times 1.46 \times 10^2$$

$$= 6.32 \times 10^{-1}$$

$$L_e = 1.46 \times 10^2 + 4.68 \times 10 + \frac{1.00}{5.36 \times 10^{-3}} \ln\left(\frac{1.00 + 6.32 \times 10^{-1}}{1 - 6.32 \times 10^{-1}} \right)$$

$$= 2.19 \times 10^2 \ \mathrm{mm} < L_a = 1.95 \times 10^3 \ \mathrm{mm}$$

$\varepsilon_{frpl,d} = \min(1.36 \times 10^{-2}, 8.70 \times 10^{-3}) = 8.70 \times 10^{-3}$

Failure is governed by intermediate debonding.

Step 2: Determine the in-plane moment capacity of the plated section. Assume the neutral axis depth $x = 1.25 \times 10^2$. With this neutral axis depth and design value of FRP limiting strain, the compressive strain at the top surface of the steel beam, $\varepsilon_{s,c}$, and the tensile strain at the bottom surface of the steel beam, $\varepsilon_{s,t}$, are found as

$$\varepsilon_{s,c} = -8.70 \times 10^{-3} \frac{-1.25 \times 10^2}{2.50 \times 10^2 + 1.00 + 1.40/2.00 - 1.25 \times 10^2} = 8.58 \times 10^{-3}$$

$$\varepsilon_{s,t} = -8.70 \times 10^{-3} \frac{2.50 \times 10^2 - 1.25 \times 10^2}{2.50 \times 10^2 + 1.00 + 1.40/2.00 - 1.25 \times 10^2}$$

$$= -8.58 \times 10^{-3}$$

The height of the beam compression yield zone can be found by using Eq. (4.3) and taking $\varepsilon_{s,i}$ as the steel yield strain.

$$1.38 \times 10^{-3} = -8.70 \times 10^{-3} \frac{h_c - 1.25 \times 10^2}{2.50 \times 10^2 + 1.00 + 1.40/2.00 - 1.25 \times 10^2}$$

$$\Rightarrow h_c = 1.05 \times 10^2 \text{ mm}$$

Therefore, the height of the elastic zone $d_e = 2.00 \times (x - h_c) = 4.00 \times 10$ mm. With these heights and using Eq. (4.4), force equilibrium can be checked:

$$\frac{f_y}{\gamma_s} t_{f1} b_{f1} + \frac{f_y}{\gamma_s} (h_c - t_{f1}) t_w + \frac{f_y}{2\gamma_s} (x - h_c) t_w - \frac{f_y}{2\gamma_s} (x - h_c) t_w$$

$$- \frac{f_y}{\gamma_s} (h - t_{f2} - 2x + h_c) t_w - \frac{f_y}{\gamma_s} t_{f2} b_{f2} - \varepsilon_{frpl,d} E_{frp} b_{frp} t_{frp}$$

$$= \frac{2.75 \times 10^2}{1.00} \times 1.18 \times 10 \times 1.66 \times 10^2 + \frac{2.75 \times 10^2}{1.00}$$

$$\times (1.25 \times 10^2 - 1.18 \times 10) \times 1.05 \times 10$$

$$+ \frac{2.75 \times 10^2}{2.00 \times 1.00} \times (1.25 \times 10^2 - 1.25 \times 10^2) - \frac{2.75 \times 10^2}{2.00 \times 1.00}$$

$$\times (1.25 \times 10^2 - 1.25 \times 10^2)$$

$$- \frac{2.75 \times 10^2}{1.00} \times (2.50 \times 10^2 - 1.18 \times 10 - 2.00 \times 1.25 \times 10^2 + 1.25 \times 10^2)$$

$$\times 1.05 \times 10 - \frac{2.75 \times 10^2}{1.00} \times 1.18 \times 10 \times 1.66 \times 10^2 - 8.70 \times 10^{-3}$$

$$\times 1.65 \times 10^5 \times 1.66 \times 10^2 \times 1.40$$

$$= -3.34 \times 10^5 \text{ N} \neq 0$$

Therefore, the force equilibrium condition is not satisfied. The neutral axis depth can now be varied until the force equilibrium criterion is satisfied. If this is done manually, it can be quite a lengthy process. However, this can be done easily using a spreadsheet such as Microsoft Excel. The neutral axis depth for this beam can be found as

$$x = 1.83 \times 10^2 \text{ mm}$$

The design moment capacity can then be found from Eq. (4.11) as

$$M_{ib,d} = 1.18 \times 10 \times 1.66 \times 10^2 \times \frac{2.75 \times 10^2}{1.00} \times \left(1.25 \times 10^2 - 5.90\right)$$

$$+ 1.60 \times 10^2 \times 1.05 \times 10 \times \frac{2.75 \times 10^2}{1.00}$$

$$\times \left(1.25 \times 10^2 - \left(1.18 \times 10 + \frac{1.60 \times 10^2}{2.00}\right)\right)$$

$$+ 1.09 \times 10 \times 1.05 \times 10 \times \frac{2.75 \times 10^2}{2.00 \times 1.00}$$

$$\times \left(1.25 \times 10^2 - \left(1.18 \times 10 + 1.60 \times 10^2 + \frac{2.00 \times 1.09 \times 10}{3.00}\right)\right)$$

$$+ 1.09 \times 10 \times 1.05 \times 10 \times - \frac{2.75 \times 10^2}{2.00 \times 1.00}$$

$$\times \left(1.25 \times 10^2 - \left(1.83 \times 10^2 + \frac{1.09 \times 10}{3.00}\right)\right)$$

$$+ 4.45 \times 10 \times 1.05 \times 10 \times - \frac{2.75 \times 10^2}{1.00}$$

$$\times \left(1.25 \times 10^2 - \left(1.83 \times 10^2 + 1.09 + \frac{4.45 \times 10}{2.00}\right)\right)$$

$$+ 1.18 \times 10 \times 1.66 \times 10^2 \times - \frac{2.75 \times 10^2}{1.00}$$

$$\times \left(1.25 \times 10^2 - \left(2.50 \times 10^2 - \frac{1.18 \times 10}{2.00}\right)\right)$$

$$+1.65 \times 10^5 \times -8.70 \times 10^{-3} \times 1.66 \times 10^2 \times 1.40$$

$$\times \left(1.25 \times 10^2 - \left(2.50 \times 10^2 + 1.00 + \frac{1.40}{2.00} \right) \right)$$

$$= 1.98 \times 10^8 \ \text{Nmm}$$

$$= 1.98 \times 10^2 \ \text{kNm}$$

4.8.3 Suppression of end debonding

End debonding needs to be avoided through the provision of FRP wraps at the laminate ends. The interfacial normal stresses can be found using either a numerical method as described in Section 4.5.4.2 or an approximate analytical method as described in Section 4.5.4.3. For this design example, the interfacial normal stresses are obtained from the analytical method of Zhang and Teng (2010b). The normalised stress distribution near the laminate end under the ultimate load is shown in Figure 4.22. Based on these results, the total peeling force to be resisted by the FRP wrap at each laminate end can be found according to Eq. (4.47):

$$P_n = 7.30 \ \text{kN}$$

It is assumed that the FRP wraps used are commercially available CFRP products with a characteristic tensile strength of 3800 MPa and a tensile elastic modulus of 235,000 MPa. The width of the FRP wrap at each plate end is taken to be equal to the width of the longitudinal FRP plate as recommended earlier, i.e., 166 mm.

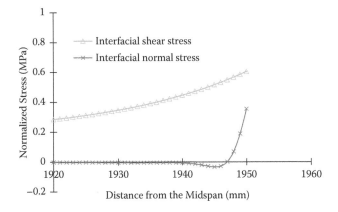

Figure 4.22 Interfacial shear and normal stress distributions at the laminate end.

The minimum FRP wrap thickness can be obtained by equating the design rupture resistance of the FRP wrap over the high peeling stress region at the plate end to the total peeling force at the plate end. For the current beam, high peeling stresses exist over a 5 mm length at the plate end.

$$P_n = 2.00 \times \frac{\sigma_{frp}}{3.00 \times \gamma_{frp}} b_{frp} t_{frp}$$

$$7.30 \times 10^3 = 2.00 \times \frac{3.80 \times 10^3}{3.00 \times 1.25} \times 5.00 \times t_{frp}$$

$$t_{frp} = 7.20 \times 10^{-1} \text{ mm}$$

Provide an FRP wrap at each plate end that meets the minimum requirements of thickness = 7.20×10^{-1} mm and width = 1.66×10^2 mm. The length of the wrap should cover the flange upper surface (for an I-section) or be extended over each side (for a rectangular section) to 200 mm above the beam soffit.

4.8.4 Design against local buckling

In an FRP-plated beam, buckling of the compression flange or web may occur even though the unplated steel section is compact. The depth of the neutral axis at the ultimate load (assuming debonding):

$$x = 1.97 \times 10^2 \text{ mm}$$

$$\lambda_{pf} = \frac{1.66 \times 10^2}{2.00 \times 1.18 \times 10} \times \sqrt{\frac{12.00 \times \left(1.00 - \left(3.00 \times 10^{-1}\right)^2\right)}{\pi^2 \times 4.30 \times 10^{-1}}} \times \sqrt{\frac{2.75 \times 10^2}{2.00 \times 10^5}}$$

$$= 4.18 \times 10^{-1}$$

$$\lambda_{pw} = \frac{1.83 \times 10^2 \times 2.00}{1.05 \times 10} \times \sqrt{\frac{12.00 \times \left(1.00 - \left(3.00 \times 10^{-1}\right)^2\right)}{\pi^2 \times 2.93}} \times \sqrt{\frac{2.75 \times 10^2}{2.00 \times 10^5}}$$

$$= 8.78 \times 10^{-1}$$

$$\lambda_s = \sqrt{4.18 \times 10^{-1} \times 8.78 \times 10^{-1}} = 6.06 \times 10^{-1}$$

$$\phi_{lb} = \frac{1.50 \times 10^{-1}}{6.06 \times 10^{-1} - 3.60 \times 10^{-1}} = 6.09 \times 10^{-1}$$

The curvature of the beam at the ultimate load:

$$\phi_x = \frac{\varepsilon_{c,x}}{x} = -8.70 \times 10^{-3} \frac{-1.00}{2.50 \times 10^2 + 1.00 + 1.40/2.00 - 1.83 \times 10^2} \times 10^3$$

$$= 1.27 \times 10^{-1}$$

$$\phi_x < \phi_{lb}$$

Therefore, compression flange buckling and web buckling are not likely to occur, and no strengthening of the compression flange or the web is required.

The likeliness of web shear buckling can be checked according to EN 1993 (2005):

$$\frac{(h - 2t_f)}{t_w} = \frac{2.50 \times 10^2 - 2.00 \times 1.18 \times 10}{1.05 \times 10} = 2.16 \times 10$$

$$72\frac{\varepsilon}{\eta} = 7.20 \times 10 \times \frac{\sqrt{\dfrac{2.35 \times 10^2}{2.75 \times 10^2}}}{1.00} = 6.45 \times 10$$

$$\frac{(h - 2t_f)}{t_w} < 72\frac{\varepsilon}{\eta} \Rightarrow OK$$

Therefore, shear buckling of the web will not occur.

4.9 CONCLUSIONS AND FUTURE RESEARCH NEEDS

This chapter has presented a systematic treatment of the flexural strengthening of steel and steel–concrete composite beams based on the available research on the topic. Within the limits of statically loaded steel I-section or RHS beams strengthened by an FRP laminate bonded to the tension flange, this chapter has provided an exposition of the possible failure modes, described methods for strength predictions, and finally, formulated a design procedure for use in practice. A design example has also been presented to illustrate the application of this design procedure.

The discussions presented in this chapter have identified many needs for further research. In the flexural strengthening of steel beams for enhancement in their static load carrying capacity, much further work is still needed to develop simple and reliable methods for a number of failure modes, such as beam lateral buckling as well as local buckling of webs and flanges. It is also necessary to conduct further research to simplify and improve some of the predictive approaches (e.g., those for debonding failures) detailed in this chapter, although these approaches are likely to be conservative. More research is also needed to facilitate the use of different strengthening schemes, such as the use of prestressed FRP laminates and the bonding of FRP laminates to locations other than the tension flange. Finally, more research is needed on the behaviour and design of FRP-strengthened steel beams subjected to fatigue and dynamic loads (earthquakes, blasts, and impacts), fire, and environmental attacks.

REFERENCES

Aoki, T., and Kubo, M. 1997. Steel members. In *Structural stability design: Steel and composite structures*, ed. Fukumoto, Y. Pergamon Press, Elsevier Science Ltd., Oxford, UK.

Al-Saidy, A.H., Klaiber, F.W., and Wipf, T.J. 2004. Repair of steel composite beams with carbon fiber-reinforced polymer plates. *Journal of Composites for Construction*, 8(2), 163–172.

Bassetti, A., Nussbaumer, A., and Hirt, M.A. 2000a. Crack repair and fatigue life extension of riveted bridge members using composite materials. Presented at Bridge Engineering Conference: Past Achievements, Current Practices, Future Technologies, Sharm El-Sheikh, Egypt.

Bassetti, A., Nussbaumer, A., and Hirt, M.A. 2000b. Fatigue life extension of riveted bridge members using prestressed carbon fiber composites. In *Proceedings, International Conference on Steel Structures of the 2000s*, Istanbul, Turkey, pp. 375–380.

Bocciarelli, M., Colombi, P., Fava, G., and Poggi, C. 2007. Interaction of interface delamination and plasticity in tensile steel members reinforced by CFRP plates. *International Journal of Fracture*, 146(1–2), 79–92.

Bocciarelli, M., Colombi, P., Fava, G., and Poggi, C. 2009. Fatigue performance of tensile steel members strengthened with CFRP plates. *Composite Structures*, 87(4), 334–343.

BS 8110. 1997. *Structural use of concrete. Part 1. Code of practice for design and construction*. London: British Standards Institution.

Chen, T., Yu, Q.Q., Gu, X.L., and Zhao, X.L. 2012. Study on fatigue behavior of strengthened non-load-carrying cruciform welded joints using carbon fiber sheets. *International Journal of Structural Stability and Dynamics*, 12(1), 179–194.

CNR-DT 202. 2005. *Guidelines for the design and construction of externally bonded FRP systems for strengthening existing structures: Metallic structures*. National Research Council, Rome, Italy.

Colombi, P., Bassetti, A., and Nussbaumer, A. 2003a. Crack growth induced delamination on steel members reinforced by prestressed composite patch. *Fatigue and Fracture of Engineering Materials and Structures*, 26(5), 429–437.

Colombi, P., Bassetti, A., and Nussbaumer, A. 2003b. Analysis of cracked steel members reinforced by pre-stress composite patch. *Fatigue and Fracture of Engineering Materials and Structures*, 26(1), 59–66.

Colombi, P., and Giulia, F. 2012. Fatigue behaviour of tensile steel/CFRP joints. *Composite Structures*, 94(8), 2407–2417.

Colombi, P., and Poggi, C. 2006. An experimental, analytical and numerical study of the static behavior of steel beams reinforced by pultruded CFRP strips. *Composites Part B: Engineering*, 37(1), 64–73.

Dawood, M., Rizkalla, S., and Sumner, E. 2007. Fatigue and overloading behavior of steel–concrete composite flexural members strengthened with high modulus CFRP materials. *Journal of Composites for Construction*, 11(6), 659–669.

De Lorenzis, L., Fernando, D., and Teng, J.G. 2013. Coupled mixed-mode cohesive zone modeling of interfacial stresses in plated beams. *International Journal of Solids and Structures*, 50 (14–15), 2477–2494.

De Lorenzis, L., and Zavarise, G. 2009. Cohesive zone modeling of interfacial stresses in plated beams. *International Journal of Solids and Structures*, 46(24), 4181–4191.

De Moura, M.F.S.F., and Chousal, J.A.G. 2006. Cohesive and continuum damage models applied to fracture characterization of bonded joints. *International Journal of Mechanical Sciences*, 48(5), 493–503.

Deng, J., and Lee, M.M.K. 2007a. Behaviour under static loading of metallic beams reinforced with a bonded CFRP plate. *Composite Structures*, 78(1), 232–242.

Deng, J., and Lee, M.M.K. 2007b. Fatigue performance of metallic beam strengthened with a bonded CFRP plate. *Composite Structures*, 78(2), 222–231.

Egilmez, O.O., and Yormaz, D. 2011. Cyclic testing of steel I-beams reinforced with GFRP. *Steel and Composite Structures*, 11, 93–114.

EN 1993 1-1. 2005. *Eurocode 3: Design of steel structures*. London: British Standards Institution.

EN 1994 1-1. 2002. *Eurocode 4: Design of composite steel and concrete structures*. London: British Standards Institution.

Fam, A., MacDougall, C., and Shaat, A. 2009. Upgrading steel–concrete composite girders and repair of damaged steel beams using bonded CFRP laminates. *Thin-Walled Structures*, 47(10), 1122–1135.

Fernando, D. 2010. Bond behaviour and debonding failures in CFRP-strengthened steel members. PhD thesis, Hong Kong Polytechnic University, Kowloon, Hong Kong.

Fernando, D., and Teng, J.G. 2013. Steel–concrete composite beams strengthened in flexure with FRP laminates, in preparation.

Fernando, D., Teng, J.G., Yu, T., and Zhao, X.L. 2012. Preparation and characterization of steel surfaces for adhesive bonding. *Journal of Composites for Construction*, in press.

Fernando, D., Yu, T., Teng, J.G., and Zhao, X.L. 2009. CFRP strengthening of rectangular steel tubes subjected to end bearing loads: Effect of adhesive properties and finite element modelling. *Thin-Walled Structures*, 47(10), 1020–1028.

Fitton, M., and Broughton, I. 2005. Variable modulus adhesives: An approach to optimised joint performance. *International Journal of Adhesion and Adhesives*, 25(4), 329–336.

Ghafoori, E., Motavalli, M., Botsis, J., Herwig, A., and Galli, M. 2012. Fatigue strengthening of damaged metallic beams using prestressed unbonded and bonded CFRP plates. *International Journal of Fatigue*, 44, 303–315.

Ghahremani, K., and Walbridge, S. 2011. Fatigue testing and analysis of peened highway bridge welds under in-service variable amplitude loading conditions. *International Journal of Fatigue*, 33, 300–312.

Goncalves, J.P.M., de Moura, M.F.S.F., Magalhaes, A.G., and de Castro, P.M.S.T. 2003. Application of interface finite elements to three-dimensional progressive failure analysis of adhesive joints. *Fatigue and Fracture of Engineering Materials and Structures*, 26(5), 479–486.

Green, S., Sause, R., and Ricles, J.M. 2002. Strength and ductility of HPS flexural members. *Journal of Constructional Steel Research*, 58(5–8), 907–941.

Haedir, J., and Zhao, X.L. 2012. Design of CFRP-strengthened steel CHS tubular beams. *Journal of Constructional Steel Research*, 72(5), 203–218.

Haghani, R., Al-Emrani, M., and Kliger, R. 2010. Effect of tapering on stress distribution in adhesive joints—Effect of tapering length and material properties. *Journal of Composite Materials*, 44(3), 287–302.

Hart-Smith, L.J. 1981. *Development in adhesives—2*. London: Applied Science Publishing.

Hollaway, L.C., Zhang, L., Photiou, N.K., Teng, J.G., and Zhang, S.S. 2006. Advances in adhesive joining of carbon fibre/polymer composites to steel members for repair and rehabilitation of bridge structures. *Advances in Structural Engineering*, 9(6), 791–803.

ICE. 2001. *FRP composites: Life extension and strengthening of metallic structures: ICE design and practice guide*, ed. Moy, S.S.J. ICE Publishing, London, UK.

Jiao, H., Zhao, X.L., and Mashiri, F. 2012. Improving fatigue performance of CFRP strengthened steel beams by applying vacuum pressure in the wet layup of CFRP woven sheets. Presented at Proceedings, Third Asia-Pacific Conference on FRP in Structures (APFIS 2012), Sapporo, Japan.

Jones, S.C., and Civjan, S.A. 2003. Application of fiber reinforced polymer overlays to extend steel fatigue life. *Journal of Composites for Construction*, 7(4), 331–338.

Kabir, M.Z., and Seif, A.E. 2010. Lateral torsional buckling of steel I-beam retrofitted using FRP sheets: Analytical solution and optimization. Presented at Proceedings, Fifth International Conference on FRP Composites in Civil Engineering (CICE 2010), Beijing, September 27–29.

Kim, Y.J., and Garrett, B. 2011. Interaction between CFRP-repair and initial damage of wide-flange steel beams subjected to three-point bending. *Composite Structures*, 93(8), 1986–1996.

Kim, Y.J., and Kent, A.H. 2012. Predictive response of notched steel beams repaired with CFRP strips including bond–slip behavior. *International Journal of Structural Stability and Dynamics*, 12(1), 1–21.

Kong, F.K., and Evans, R.H. 1987. *Reinforced and prestressed concrete*. 3rd ed. London: Chapman & Hall.

Lenwari, A., Thepchatri, T., and Albrecht, P. 2006. Debonding strength of steel beams strengthened with CFRP plates. *Journal of Composites for Construction*, 10(1), 69–78.

Li, S., Thouless, M.D., Waas, A.M., Schroeder, J.A., and Zavattieri, P.D. 2005. Use of a cohesive-zone model to analyze the fracture of a fiber-reinforced polymer-matrix composite. *Composites Science and Technology*, 65(3–4), 537–549.

Li, S., Thouless, M.D., Waas, A.M., Schroeder, J.A., and Zavattieri, P.D. 2006. Mixed-mode cohesive-zone models for fracture of an adhesively bonded polymer-matrix composite. *Engineering Fracture Mechanics*, 73(1), 64–78.

Liljedahl, C.D.M., Crocombe, A.D., Wahab, M.A., and Ashcroft, I.A. 2006. Damage modelling of adhesively bonded joints. *International Journal of Fracture*, 141(1–2), 147–161.

Linghoff, D., and Al-Emrani, M. 2010. Performance of steel beams strengthened with CFRP laminate. Part 2. FE analysis. *Composites Part B: Engineering*, 41(7), 516–522.

Linghoff, D., Haghani, R., and Al-Emrani, M. 2009. Carbon-fibre composites for strengthening steel structures. *Thin-Walled Structures*, 47(10), 1048–1058.

Liu, H., Al-Mahaidi, R., and Zhao, X.L. 2009a. Experimental study of fatigue crack growth behavior in adhesively reinforced steel structures. *Composite Structures*, 90(1), 12–20.

Liu, H., Xiao, Z., Zhao, X.L., and Al-Mahaidi, R. 2009b. Prediction of fatigue life for CFRP-strengthened steel plates. *Thin-Walled Structures*, 47(10), 1069–1077.

Maddox, S.J. 1985. Improving the fatigue strength of welded joints by peening. *Metal Construction*, 17, 220–224.

Megueni, A., Bouiadjra, B.B., and Belhouari, M. 2004. Disbond effect on the stress intensity factor for repairing cracks with bonded composite patch. *Computational Material Science*, 29(4), 407–413.

Miller, T.C., Chajes, M.J., Mertz, D.R., and Hastings, J.N. 2001. Strengthening of a steel bridge girder using CFRP plates. *Journal of Bridge Engineering*, 6(6), 523–528.

Nakamura, H., Jiang, W., Suzuki, H., Maeda, K., and Irube, T. 2009. Experimental study on repair of fatigue cracks at welded web gusset joint using CFRP strips. *Thin-Walled Structures*, 47(10), 1059–1068.

Nakamura, H., Maeda, K., Suzuki, H., and Irube, T. 2008. Monitoring for fatigue crack propagation of steel plate repaired by CFRP strips. In *Proceedings, Fourth International Conference on Bridge Maintenance, Safety and Management, IABMAS '08*, pp. 2943–2950.

Nozaka, K., Shield, C.K., and Hajjar, J.F. 2005. Effective bond length of carbon-fiber-reinforced polymer strips bonded to fatigued steel bridge I-girders. *Journal of Bridge Engineering*, 10(2), 195–205.

Okeil, A.M., Bingol, Y., and Ferdous, Md.R. 2009. Novel technique for inhibiting buckling of thin-walled steel structures using pultruted glass FRP sections. *Journal of Composites for Construction*, 13(6), 547–557.

Ouinas, D., Serier, B., and Bouiadjra, B.B. 2007. The effects of disbonds on the pure mode II stress intensity factor of aluminum plate reinforced with bonded composite materials. *Computational Material Science*, 39(4), 782–787.

Pi, Y.L., and Trahair, N.S. 1994. Inelastic bending and torsion of steel I-beams. *Journal of Structural Engineering*, 120(12), 3397–3417.

Pi, Y.L., and Trahair, N.S. 1995. Inelastic torsion of steel I-beams. *Journal of Structural Engineering*, 121(4), 609–620.

Rabinovitch, O., and Frostig, Y. 2000. Closed-form high-order analysis of RC beams strengthened with FRP strips. *Journal of Composites for Construction*, 4(2), 65–74.

Roberts, T.M. 1989. Approximate analysis of shear and normal stress concentrations in the adhesive layer of plated RC beams. *Structural Engineering*, 67(12), 229–233.

Saadatmanesh, H., and Malek, A.H. 1998. Design guidelines for flexural strengthening of RC beams with FRP plates. *Journal of Composites for Construction*, 2(4), 158–164.

Sallam, H.E.M., Ahmad, S.S.E., Badawy, A.A.M., and Mamdouh, W. 2006. Evaluation of steel I-beams strengthened by various plating methods. *Advances in Structural Engineering*, 9(4), 535–544.

Sayed-Ahmed, E.Y. 2006. Numerical investigation into strengthening steel I-section beams using CFRP strips. In *ASCE Structures Congress*, pp. 1–8.

Schnerch, D., Dawood, M., Rizkalla, S., and Sumner, E. 2007. Proposed design guidelines for strengthening of steel bridges with FRP materials. *Construction Building Materials*, 21(5), 1001–1010.

Schnerch, D., and Rizkalla, S. 2008. Flexural strengthening of steel bridges with high modulus CFRP strips. *Journal of Bridge Engineering*, ASCE, 13(2), 192–201.

Seica, M., and Packer, J. 2007. FRP materials for the rehabilitation of tubular steel structures for underwater applications. *Composite Structures*, 80(3), 440–450.

Sen, R., Liby, L., and Mullins, G. 2001. Strengthening steel bridge sections using CFRP laminates. *Composites Part B: Engineering*, 32(4), 309–322.

Shaat, A., and Fam, A. 2008. Repair of cracked steel girders connected to concrete slabs using carbon-fiber-reinforced polymer sheets. *Journal of Composites for Construction*, 12(6), 650–659.

Shen, H.S., Teng, J.G., and Yang, J. 2001. Interfacial stresses in beams and slabs bonded with thin plate. *Journal of Engineering Mechanics*, 127(4), 399–406.

Smith, S.T., and Teng, J.G. 2001. Interfacial stresses in plated beams. *Engineering Structures*, 23(7), 857–871.

Sorensen, B.F. 2002. Cohesive law and notch sensitivity of adhesive joints. *Acta Materialia*, 50(5), 1053–1061.

Sorensen, B.F., and Jacobsen, T.K. 2003. Determination of cohesive laws by the J integral approach. *Engineering Fracture Mechanics*, 70(14), 1841–1858.

Stratford, T., and Cadei, J. 2006. Elastic analysis of adhesion stresses for the design of a strengthening plate bonded to a beam. *Construction Building Materials*, 20(1–2), 34–45.

Taljsten, B. 1997. Strengthening of beams by plate bonding. *Journal of Materials in Civil Engineering*, 9(4), 206–212.

Taljsten, B., Hansen, C.S., and Schmidt, J.W. 2009. Strengthening of old metallic structures in fatigue with prestressed and non-prestressed CFRP laminates. *Construction Building Materials*, 23(4), 1665–1677.

Tavakkolizadeh, M., and Saadatmanesh, H. 2003a. Strengthening of steel–concrete composite girders using carbon fiber reinforced polymers sheets. *Journal of Structural Engineering*, 129(1), 30–40.

Tavakkolizadeh, M., and Saadatmanesh, H. 2003b. Fatigue strength of steel girders strengthened with carbon fiber reinforced polymer path. *Journal of Structural Engineering*. ASCE, 129(2), 186–196.

Teng, J.G., Chen, J.F., Smith, S.T., and Lam, L. 2002a. *FRP strengthened RC structures*. Chichester, UK: John Wiley & Sons.

Teng, J.G., Smith, S.T., Yao, J., and Chen, J.F. 2003. Intermediate crack-induced debonding in RC beams and slabs. *Construction and Building Materials*, 17(6–7), 447–462.

Teng, J.G., Yu, T., and Fernando, D. 2012. Strengthening of steel structures with FRP composites. *Journal of Constructional Steel Research*, 78, 131–143.

Teng, J.G., Yuan, H., and Chen, J.F. 2006. FRP-to-concrete interfaces between two adjacent cracks: Theoretical model for debonding failure. *International Journal of Solids and Structures*, 43(18–19), 5750–5778.

Teng, J.G., Zhang, J.W., and Smith, S.T. 2002b. Interfacial stresses in reinforced concrete beams bonded with a soffit plate: A finite element study. *Construction Building Materials*, 16(1), 1–14.

Trahair, N.S. 1993. *Flexural-torsional buckling of structures*. Boca Raton, FL: CRC Press.

Tsouvalis, N.G., Mirisiotis, L.S., and Dimou, D.N. 2009. Experimental and numerical study of the fatigue behaviour of composite patch reinforced cracked steel plates. *International Journal of Fatigue*, 31(10), 1613–1627.

Vasdravellis, G., Uy, B., Tan, E.L., and Kirkland, B. 2012. Behavior and design of composite beams subjected to negative bending and compression. *Journal of Constructional Steel Research*, 79, 34–47.

Wu, C., Zhao, X.L., Al-Mahaidi, R., Emdad, M., and Duan, W.H. 2012. Fatigue tests of cracked steel plates strengthened with UHM CFRP plates. *Advances in Structural Engineering*, 15(10), 1801–1815.

Xiao, Z.G., and Zhao, X.L. 2012. CFRP repaired welded thin-walled cross-beam connections subject to in-plane fatigue loading. *International Journal of Structural Stability and Dynamics*, 12(1), 195–211.

Yang, J., and Ye, J.Q. 2010. An improved closed-form solution to interfacial stresses in plated beams using a two-stage approach. *International Journal of Mechanical Sciences*, 52(1), 13–30.

Yu, T., Fernando, D., Teng, J.G., and Zhao, X.L. 2012. Experimental study on CFRP-to-steel bonded interfaces. *Composites Part B: Engineering*, 43(5), 2279–2289.

Yuan, H., and Xu, Y. 2008. Computational fracture mechanics assessment of adhesive joints. *Computational Materials Science*, 43(1), 146–156.

Zhang, L., Hollaway, L.C., Teng, J.G., and Zhang, S.S. 2006. Strengthening of steel bridges under low frequency vibrations. Presented at Proceedings, Third International Conference on FRP Composites in Civil Engineering, Miami, FL, December 13–15.

Zhang, L., and Teng, J.G. 2007. Elastic lateral buckling of I-section beams bonded with a thin plate. In *Proceedings, Third International Conference on Advanced Composites in Construction (ACIC 2007)*, University of Bath, UK, April 2–4, 2007, pp. 441–446.

Zhang, L., and Teng, J.G. 2010a. Finite element prediction of interfacial stresses in structural members bonded with a thin plate. *Engineering Structures*, 32(2), 459–471.

Zhang, L., and Teng, J. 2010b. Simple general solution for interfacial stresses in plated beams. *Journal of Composites for Construction*, 14(4), 434–442.

Zhao, X.L., Fernando, D., and Al-Mahaidi, R. 2006. CFRP strengthened RHS subjected to transverse end bearing force. *Engineering Structures*, 28(11), 1555–1565.

Chapter 5

Strengthening of compression members

5.1 GENERAL

Steel members under compression often fail in local or flexural–torsional collapse modes (distortional failures may also occur for open sections), depending on the cross section and column slenderness values. The presence of fibre-reinforced polymer (FRP) can delay, eliminate, or reduce the buckling/collapse, leading to a higher load carrying capacity. The improved performance of compression members due to FRP strengthening is summarised in Table 5.1, for steel members with circular hollow sections (CHSs), square hollow sections (SHSs), lipped channel sections, and T-sections.

This chapter first addresses the methods of FRP strengthening steel compression members and then the corresponding structural behaviour, expressed in terms of failure modes and load versus displacement curves. Then the section capacity formulae are presented for each type of section (CHS, SHS, lipped channel, and T-section), followed by the member capacity formulae for SHS and lipped channel sections only. Due to lack of research, no member capacity formulae are yet available for CHS and T-sections. Finally, design examples are given for carbon fibre-reinforced polymer (CFRP)-strengthened CHS and SHS columns.

5.2 METHODS OF STRENGTHENING

For the strengthening of SHS columns, both longitudinal and transverse FRPs can be utilised as shown in Figure 5.1(a). For example, the strengthening scheme 1T1L implies that one layer of CFRP is applied in the transverse direction (i.e., the fibre direction is perpendicular to the column axis), followed by a layer of CFRP in the longitudinal direction (i.e., the fibre direction is parallel to the column axis). More details of the strengthening procedure can be found in Bambach and Elchalakani (2007). A similar approach (i.e., using a combination of longitudinal and transverse CFRP) was adopted by Haedir and Zhao (2011) to strengthen CHS columns.

Table 5.1 Examples of improved performance of compression members due to FRP strengthening

FRP used (Young's modulus)	Type of compression members	Improved performance	Reference
CFRP sheet (230 GPa)	CHS stub columns (diameter-to-thickness ratio from 37 to 78)	Compression strength increases up to 75%	Haedir and Zhao (2011)
GFRP sheet (76 GPa)	CHS stub columns (diameter-to-thickness ratio of 39)	Compression strength increases up to 10%	Teng and Hu (2007)
CFRP sheet (115 or 230 GPa)	SHS stub columns (width-to-thickness ratio of 28)	Compression strength increases up to 18%	Shaat and Fam (2006), Shaat (2007)
CFRP sheet (230 GPa)	SHS stub columns (width-to-thickness ratio from 50 to 120)	Compression strength increases up to 2.8 times	Bambach et al. (2009a)
CFRP sheet (230 GPa)	SHS slender columns (width-to-thickness ratio from 14 to 28)	Compression strength increases up to 23%	Shaat and Fam (2006, 2009)
CFRP sheet (235 GPa)	Lipped channel stub columns (web depth-to-thickness ratio of 126)	Compression strength increases up to 15%	Silvestre et al. (2008)
CFRP sheet (235 GPa)	Lipped channel slender columns (web depth-to-thickness ratio of 81)	Compression strength increases up to about 20%	Silvestre et al. (2008)
CFRP sheet (235 GPa) and GFRP sheet (41 GPa)	T-section stub column (web depth-to-thickness ratio of 28)	Compression strength increases up to 14%	Harries et al. (2009)
CFRP sheet (235 GPa) and GFRP sheet (41 GPa)	T-section slender column (web depth-to-thickness ratio of 28)	Compression strength increases up to 9%	Harries et al. (2009)
CFRP sheet (230 GPa)	SHS stub columns under large axial deformation (width-to-thickness ratio from 25 to 50)	Compression strength increases up to 102%; energy absorption increases up to 113%	Bambach and Elchalakani (2007)

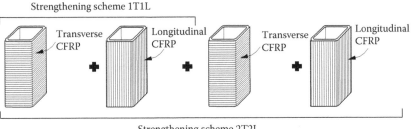

Strengthening scheme 1T1L

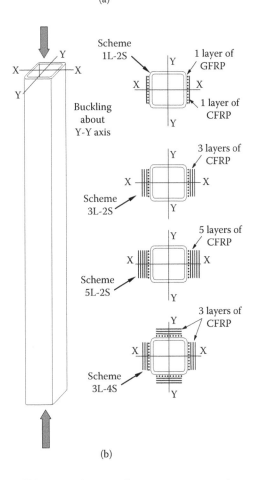

Figure 5.1 Examples of FRP strengthening of compression members. (a) SHS stub column. (From Bambach, M.R. et al., *Composite Structures*, 87(3), 282–292, 2009.) (b) SHS slender column. (Adapted from Shaat, A., and Fam, A., *Canadian Journal of Civil Engineering*, 33(4), 458–470, 2006; Shaat, A., and Fam, A., *Journal of Composites for Construction*, 13(1), 2–12, 2009.)

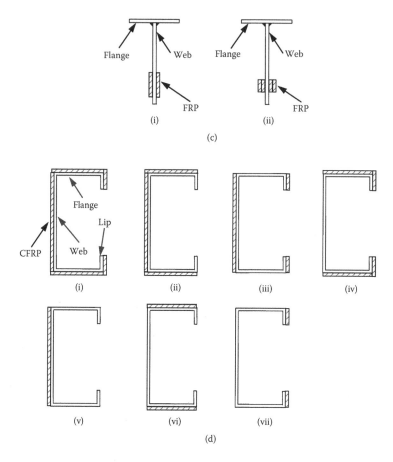

Figure 5.1 (Continued) Examples of FRP strengthening of compression members. (c) T-section: (i) one layer of wider FRP sheet and (ii) two layers of narrower FRP sheets. (Adapted from Harries, K.A. et al. *Thin-Walled Structures*, 47(10), 1092–1101, 2009.) (d) Lipped channel section: (i) scheme WFL, (ii) scheme WF, (iii) scheme WL, (iv) scheme FL, (v) scheme W, (vi) scheme F, and (vii) scheme L. (Adapted from Silvestre, N. et al., *Engineering Structures*, 30(10), 2613–2630, 2008.)

Shaat and Fam (2006, 2009) applied FRP only to the flat faces (two or four) of SHS, as shown in Figure 5.1(b). Furthermore, one layer of GFRP was installed on the steel surface to prevent direct contact between carbon fibre and steel, which could lead to galvanic corrosion. The number of CFRP layers varied from one to five on two or four flat faces of the SHS. For example, the strengthening scheme 5L2S refers to five layers of CFRP applied on two flat faces.

For the strengthening of a T-section, FRP (either GFRP or CFRP) was applied by Harries et al. (2009) on the more critical element, i.e., web, as shown in Figure 5.1(c). Two options were explored: to use (1) one layer of wider FRP or (2) two layers of narrower FRP.

For the strengthening of a lipped channel section, CFRP can be applied to individual walls (web (W), flanges (F), or lips (L)) or a combination of the elements. Seven strengthening schemes are shown in Figure 5.1(d). The CFRP fibre orientation is either longitudinal (L), i.e., parallel to the column axis, or transverse (T), i.e., perpendicular to the column axis. For example, the strengthening scheme WF-L implies that CFRP is applied on the web and flanges, and that the fibre is parallel to the column axis.

5.3 STRUCTURAL BEHAVIOUR

5.3.1 Failure modes

The failure mode of CHS under compression varies with the diameter-to-thickness ratio (d_s/t_s). For relatively thick sections, an "elephant foot" failure mode (i.e., inelastic local buckling) is often found (e.g., Teng and Hu 2007), as shown in Figure 5.2(a)(i). Elastic local buckling often occurs for a relatively thin section (e.g., Haedir and Zhao 2011), as shown in Figure 5.2(a)(ii). Note also that CFRP strengthening can delay or eliminate local buckling of CHS, as demonstrated in Figure 5.2(a).

Three typical failure modes were identified by Shaat and Fam (2006) for short SHS columns strengthened by CFRP, as schematically shown in Figure 5.2(b). They are (1) delamination between steel and the longitudinal CFRP at the end of the specimen, (2) rupture of transverse CFRP near the corners, and (3) delamination between steel and the transverse CFRP at an inward buckling location.

For slender SHS columns (bending about the y-y axis, as shown in Figure 5.1(b)) without CFRP, it is well known that failure is mainly due to excessive overall buckling, followed by a secondary local buckling near the mid-length of the column. The local buckling is in the form of inward buckling of the compression face and outward buckling of the two side faces. For the CFRP-strengthened specimens (see Figure 5.2(c)), delamination and crushing of CFRP were observed on the compression side (where inward buckling occurred). CFRP rupture was found at the outward buckling (tension side of the buckled column) location. Shaat and Fam (2009) also found that CFRP debonding can occur before or after the overall buckling of the column, depending on the column slenderness and the amount of CFRP applied.

For lipped channel section columns without CFRP strengthening, three commonly observed failures are associated with local, distortional, and flexural–torsional modes (Hancock 2007). When CFRP is applied the failure mode was found (Silvestre et al. 2008) to be caused by interaction between local and distortional buckling. As the load increases, such interaction causes excessive column wall deformation, which leads

CHS without GFRP   GFRP strengthened CHS

(i) GFRP strengthened CHS with d_s/t_s of 28. (Courtesy of J.G. Teng, The Hong Kong Polytechnic University, China.)

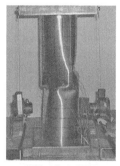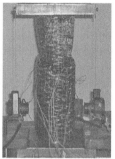

CHS without CFRP   CFRP strengthened CHS

(ii) CFRP strengthened CHS with d_s/t_s of 75. (Courtesy of J. Haedir, Monash University, Australia.)

(a)

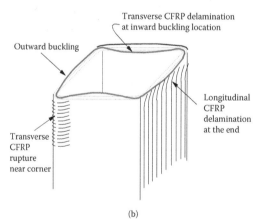

(b)

Figure 5.2 Examples of failure modes. (a) CHS strengthened by FRP. (b) SHS strengthened by FRP. (From Zhao, X.L., and Zhang, L., *Engineering Structures*, 29(8), 1808–1823, 2007.)

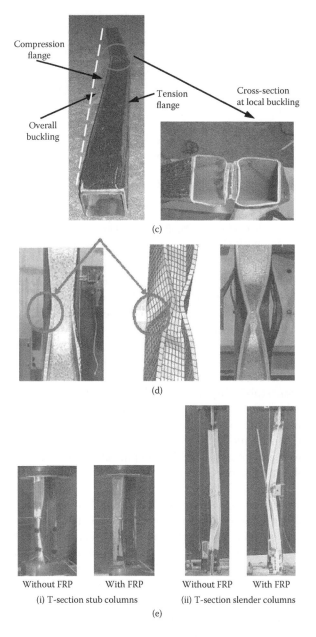

Figure 5.2 (Continued) Examples of failure modes. (c) SHS slender columns. (Courtesy of A. Fam, Queens University, Canada.) (d) Channel section columns. (Courtesy of B. Young, University of Hong Kong, China.) (e) T-section columns. (Courtesy of K. Harries, University of Pittsburgh.)

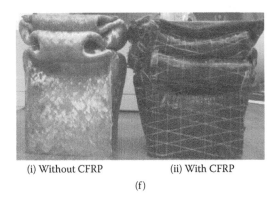

(i) Without CFRP   (ii) With CFRP

(f)

Figure 5.2 (Continued) Examples of failure modes. (f) SHS under large deformation. (Courtesy of M. Bambach, University of New South Wales, Australia.)

to CFRP debonding. This often occurs after the peak load is reached. Figure 5.2(d) shows the initial CFRP debonding at the yielded zone and the total debonding of the CFRP sheets. Strengthening only the flanges or the web leads to quite marginal ultimate load increases. It is preferable to strengthen both the flanges and the web. Moreover, the strengthening of the lips does not provide a meaningful increase in load carrying capacity.

For T-section stub columns strengthened by CFRP, little local buckling deformation was observed, as shown in Figure 5.2(e)(i). For T-section slender columns with or without FRP strengthening, flexural–torsional is the typical failure mode, as shown in Figure 5.2(e)(ii). A plastic "kinking" of the web was observed (Harries et al. 2009) for an unstrengthened T-section column. The presence of FRP mitigated this post-buckling crippling effect.

For SHS stub columns without CFRP strengthening under large deformation, the cross sections deform in an axis-symmetric collapse mode, with the flat sides of the SHS forming the roof mechanism shown in Figure 5.2(f) (i). When CFRP was applied, the SHS stub column failure mode was found (Bambach and Elchalakani 2007) to be very similar to that of unstrengthened SHS, i.e., involving multiple folding mechanisms. No delamination occurred prior to the attainment of the ultimate load, which indicates that the CFRP was well bonded to the SHS. During the large deformation crushing process, the CFRP typically delaminated from the steel at the column ends, undergoing rupture at the corners and at the exterior of the folds (see Figure 5.2(f)(ii)). For the more slender specimens, large deformation crushing also produced delamination of the CFRP across some of the folds, but little rupture of the CFRP was observed.

5.3.2 Load versus displacement curves

The behaviour of FRP-strengthened steel members can be assessed using the load versus displacement curves obtained from experimental studies. Typical curves are shown in Figure 5.3 for various types of sections with FRP strengthening and the strengthening schemes defined in Section 5.2. The general trend is clear: FRP increases the load carrying capacity, ductility, and energy absorption. The improvement due to FRP strengthening is also summarised in Table 5.1.

5.4 CAPACITY OF FRP-STRENGTHENED STEEL SECTIONS

5.4.1 CFRP-strengthened CHS sections

Haedir and Zhao (2011) carried out a study on CFRP-strengthened CHS in compression. Design formulae were proposed and related to the existing standards (AS 4100 (Standards Australia 1988) and EC3-Part 1.1 (CEN 2005)) for the design of CHS stub columns. Section 5.4.1 is based on the work of Haedir and Zhao (2011).

5.4.1.1 Modified AS 4100 model

The element slenderness (λ_s) for CHS in AS 4100 is defined as

$$\lambda_s = \left(\frac{d_s}{t_s}\right)\left(\frac{\sigma_{y,s}}{250}\right) \tag{5.1}$$

where d_s, t_s, and $\sigma_{y,s}$ are the CHS outside diameter, thickness, and yield stress.

When CFRP is applied to a CHS, the element slenderness for the equivalent CHS (λ_{es}) can be defined in the same format as

$$\lambda_{es} = \left(\frac{d_{es}}{t_{es}}\right)\left(\frac{\sigma_{y,s}}{250}\right) \tag{5.2}$$

in which d_{es} is the outer diameter of the equivalent section, and t_{es} is the total thickness of the steel and supplanted sections, defined by

$$d_{es} = d_s + 2t_{es,cs} \tag{5.3}$$

$$t_{es} = t_s + t_{es,cs} \tag{5.4}$$

where $t_{es,cs}$ is the thickness due to CFRP strengthening in both the longitudinal and transverse directions. It can be estimated by

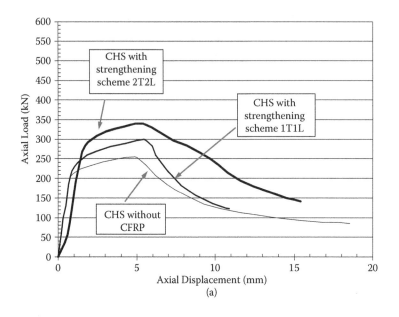

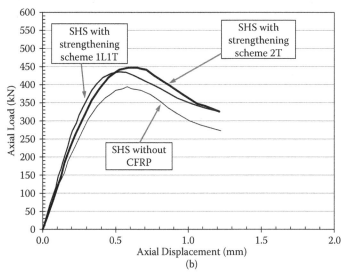

Figure 5.3 Typical load versus displacement curves. (a) CHS stub columns with d_s/t_s of 38. (Adapted from Haedir, J., and Zhao, X.L., *Journal of Constructional Steel Research*, 67(3), 497–509, 2011.) (b) SHS stub columns with b/t_s of 28. (Adapted from Shaat, A., and Fam, A., *Canadian Journal of Civil Engineering*, 33(4), 458–470, 2006.)

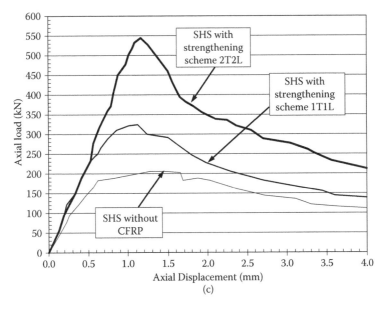

(c)

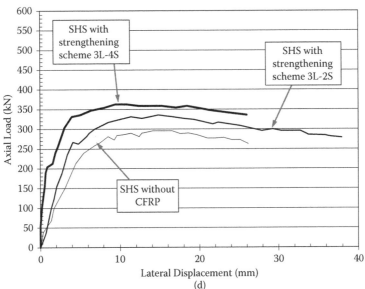

(d)

Figure 5.3 (Continued) Typical load versus displacement curves. (c) SHS stub columns with b/t$_s$ of 90. (Adapted from Bambach. M.R. et al., *Thin-Walled Structures*, 47(10), 1112–1121, 2009.) (d) SHS slender columns with b/t$_s$ of 28. (Adapted from Shaat, A., and Fam, A., *Canadian Journal of Civil Engineering*, 33(4), 458–470, 2006.)

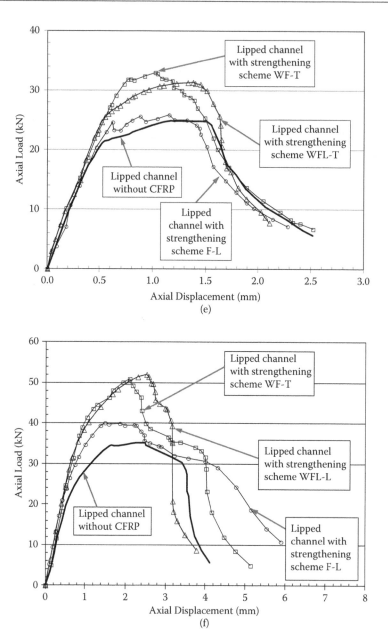

Figure 5.3 (Continued) Typical load versus displacement curves. (e) Lipped channel section stub columns with d/t_s of 126. (Adapted from Silvestre, N. et al., *Engineering Structures*, 30(10), 2613–2630, 2008.) (f) Lipped channel section slender columns with d/t_s of 81. (Adapted from Silvestre, N. et al., *Engineering Structures*, 30(10), 2613–2630, 2008.)

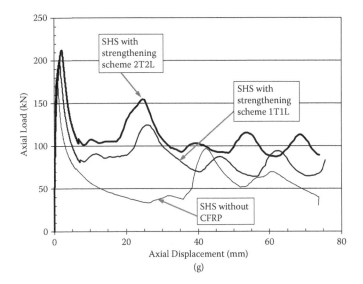

Figure 5.3 (Continued) Typical load versus displacement curves. (g) SHS short columns with b/t$_s$ of 25 under large axial deformation. (Adapted from Bambach, M.R., and Elchalakani, M., *Thin-Walled Structures*, 45(2), 159–170, 2007.)

$$t_{es,cs} = n_L \beta_L t_{L,cs} + n_T \beta_T t_{T,cs} \tag{5.5}$$

in which (1) $t_{L,cs}$ and $t_{T,cs}$ are the thicknesses of the longitudinal and transverse carbon fibre sheets, and (2) β_L and β_T are the modular ratios associated with the longitudinal and transverse fibres, defined as

$$\beta_L = \frac{E_{L,t,cs}}{E_s} = \frac{E_{L,c,cs}}{E_s} \tag{5.6}$$

$$\beta_T = \frac{E_{T,t,cs}}{E_s} = \frac{E_{T,c,cs}}{E_s} \tag{5.7}$$

where (1) $E_{L,t,cs}$ and $E_{L,c,cs}$ are Young's moduli of the longitudinal carbon fibres in tension and compression, and (2) $E_{T,t,cs}$ and $E_{T,c,cs}$ are Young's moduli of the transverse carbon fibres in tension and compression. For simplicity, the moduli of elasticity in tension and compression are assumed to be the same.

The proportioning factor ξ, reflecting the modulus ratio between transverse and longitudinal directions, is given by

$$\xi = \frac{\beta_T}{\beta_L} = \frac{E_{T,t,cs}}{E_{L,t,cs}} \tag{5.8}$$

The thickness of the carbon fibres is assumed to be the same in longitudinal and transverse directions, i.e.,

$$t_{L,cs} = t_{T,cs} = t_{CFRP\_sheet} \tag{5.9}$$

Substituting Equations (5.6) to (5.9) into (5.5) yields

$$t_{es,cs} = \beta_L t_{CFRP\_sheet} (n_L + \xi n_T) \tag{5.10}$$

A nondimensional fibre reinforcement factor (α) representing, to some extent, the amount of CFRP applied is defined as

$$\alpha = \frac{(n_L + n_T)t_{CFRP\_sheet}}{t_s} \tag{5.11}$$

The section capacity of the CFRP-reinforced CHS column under axial compression is determined as

$$N_{s,CHS\ with\ CFRP,AS4100} = A_{eff,\ eq}\sigma_{y,\ s} \tag{5.12}$$

where $A_{eff,eq}$ is the effective area of the equivalent steel section. The effective diameter of the equivalent steel section $d_{eff,eq}$, due to local buckling, can be written in a similar format in AS 4100 as

$$d_{eff,eq} = d_{es}\sqrt{\frac{82}{\lambda_{es}}} \tag{5.13}$$

Nondimensionalising $N_{s,CHS\ with\ CFRP,AS4100}$ with respect to the yield capacity of CHS ($N_{y,s}$) gives

$$\frac{N_{s,CHS\ with\ CFRP,AS4100}}{N_{y,s}} = \frac{A_{eff,eq}}{A_s} \tag{5.14}$$

where A_s is the cross-sectional area of the CHS.

The above expression can be expanded in the following form:

1. For $\dfrac{d_{es}}{t_{es}} \leq \dfrac{82}{\left(\dfrac{\sigma_{y,s}}{250}\right)}$

$$\frac{A_{eff,eq}}{A_s} \approx \left(1 + \frac{t_{es,cs}}{t_s}\right) \tag{5.15a}$$

2. For $\dfrac{d_{es}}{t_{es}} > \dfrac{82}{\left(\dfrac{\sigma_{y,s}}{250}\right)}$

$$\dfrac{A_{eff,eq}}{A_s} \approx \left(1 + \dfrac{t_{es,cs}}{t_s}\right)\sqrt{\dfrac{82}{\left(\dfrac{d_s}{t_s}\right)\left(\dfrac{\sigma_{y,s}}{250}\right)}\left(1 + \dfrac{t_{es,cs}}{t_s}\right)} \qquad (5.15b)$$

Hence:

1. For $\dfrac{d_{es}}{t_{es}} \leq \dfrac{82}{\left(\dfrac{\sigma_{y,s}}{250}\right)}$

$$N_{s,\text{CHS with CFRP,AS4100}} \approx \left(1 + \dfrac{t_{es,cs}}{t_s}\right) \cdot A_s\sigma_{y,s} \qquad (5.16a)$$

2. For $\dfrac{d_{es}}{t_{es}} > \dfrac{82}{\left(\dfrac{\sigma_{y,s}}{250}\right)}$

$N_{s,\text{CHS with CFRP,AS4100}}$

$$\approx \left(1 + \dfrac{t_{es,cs}}{t_s}\right)\sqrt{\dfrac{82}{\left(\dfrac{d_s}{t_s}\right)\left(\dfrac{\sigma_{y,s}}{250}\right)}\left(1 + \dfrac{t_{es,cs}}{t_s}\right)} \cdot A_s\sigma_{y,s} \qquad (5.16b)$$

where $\sigma_{y,s}$ is the CHS yield stress in MPa.

The validity range of the parameters appearing in the modified AS 4100 model is as follows:

$37 \leq d_s/t_s \leq 78$

$250\ \text{MPa} \leq \sigma_{y,s} \leq 450\ \text{MPa}$

$0.1 \leq \alpha \leq 0.3$

5.4.1.2 Modified EC3 model

Similar to the modified AS 4100 model, the section capacity of CFRP-strengthened CHS columns can be expressed as

$$N_{s,\text{CHS with CFRP,EC3}} = A_{eff,eq}\sigma_{y,s} \qquad (5.17)$$

where the effective area of the equivalent steel section $A_{eff,eq}$ can be estimated by

$$A_{eff,eq} = \rho_{es}A_{es} \tag{5.18}$$

in which A_{es} is the cross-sectional area of the equivalent steel section (i.e., CHS with d_{es} and t_{es} defined in Equations (5.3) and (5.4)) without any reduction due to local buckling, and ρ_{es} is taken as the commonly used winter reduction factor calculated using the dimensions of the equivalent steel section. Then,

$$A_{es} = \pi\left(t_s + t_{es,cs}\right)^2 \left(\frac{\dfrac{d_s}{t_s} + 2\dfrac{t_{es,cs}}{t_s}}{1 + \dfrac{t_{es,cs}}{t_s}} - 1 \right) \tag{5.19}$$

where $t_{es,cs}$ is given in Equation (5.5), and

$$\rho_{es} = \frac{\lambda_{esp} - 0.22}{\lambda_{esp}^2} \leq 1.0 \tag{5.20}$$

$$\lambda_{esp} = \frac{\sigma_{y,s}}{\sigma_{ese}} \leq 1.0 \tag{5.21}$$

where σ_{ese} is the elastic buckling stress according to Plantema's expression:

$$\sigma_{ese} = \frac{0.265E_s}{\sqrt{3}\sqrt{1-v^2}}\left(\frac{t_{es}}{d_{es}}\right) \tag{5.22}$$

Substituting Equation (5.22) into Equation (5.21), with $E_s = 200,000$ MPa and $v = 0.3$, leads to

$$\lambda_{esp} = \frac{\left(\dfrac{\dfrac{d_s}{t_s} + 2\dfrac{t_{es,cs}}{t_s}}{1 + \dfrac{t_{es,cs}}{t_s}} \right)\left(\dfrac{\sigma_{y,s}}{235} \right)}{137} \leq 1.0 \tag{5.23}$$

Nondimensionalising $N_{s,CHS\ with\ CFRP,Ec3}$ with respect to the yield capacity of CHS ($N_{y,s}$),

$$\frac{N_{s,CHS\ with\ CFRP,EC3}}{N_{y,s}} = \frac{A_{eff,eq}}{A_s} \tag{5.24}$$

where $A_{eff,eq}$ is given in Equation (5.18) and A_s is the cross-sectional area of the CHS.

The nondimensional ratio in Equation (5.24) can be expressed for different classes of CHS defined in EC3 as follows:

1. For class 1, 2, and 3 cross sections with $\dfrac{d_{es}}{t_{es}} \leq \dfrac{90}{\left(\dfrac{\sigma_{y,s}}{235}\right)}$:

$$\frac{A_{eff,eq}}{A_s} \approx \left(1 + \frac{t_{es,cs}}{t_s}\right) \tag{5.25a}$$

2. For class 4 cross sections with $\dfrac{d_{es}}{t_{es}} > \dfrac{90}{\left(\dfrac{\sigma_{y,s}}{235}\right)}$:

$$\frac{A_{eff,eq}}{A_s} \approx \left(1 + \frac{t_{es,cs}}{t_s}\right)^2 \cdot \frac{137\left(\dfrac{\sigma_{y,s}}{235}\right)\left(\dfrac{d_s}{t_s}\right) - 30\left(1 + \dfrac{t_{es,cs}}{t_s}\right)}{\left(\dfrac{\sigma_{y,s}}{235}\right)^2\left(\dfrac{d_s}{t_s}\right)^2} \tag{5.25b}$$

Hence,

1. For class 1, 2, and 3 cross sections with $\dfrac{d_{es}}{t_{es}} \leq \dfrac{90}{\left(\dfrac{\sigma_{y,s}}{235}\right)}$:

$$N_{s,CHS\ with\ CFRP,EC3} \approx \left(1 + \frac{t_{es,cs}}{t_s}\right) \cdot A_s\sigma_{y,s} \tag{5.26a}$$

2. For class 4 cross sections with $\dfrac{d_{es}}{t_{es}} > \dfrac{90}{\left(\dfrac{\sigma_{y,s}}{235}\right)}$:

$N_{s,CHS\ with\ CFRP,EC3}$

$$\approx \left(1 + \frac{t_{es,cs}}{t_s}\right)^2 \cdot \frac{137\left(\dfrac{\sigma_{y,s}}{235}\right)\left(\dfrac{d_s}{t_s}\right) - 30\left(1 + \dfrac{t_{es,cs}}{t_s}\right)}{\left(\dfrac{\sigma_{y,s}}{235}\right)^2\left(\dfrac{d_s}{t_s}\right)^2} \cdot A_s\sigma_{y,s} \tag{5.26b}$$

where $\sigma_{y,s}$ is the yield stress of CHS in MPa.

The validity range of the parameters appearing in the modified EC3 model is as follows:

$$37 \leq d_s/t_s \leq 78$$

$$235 \text{ MPa} \leq \sigma_{y,s} \leq 460 \text{ MPa}$$

$$0.1 \leq \alpha \leq 0.3$$

5.4.1.3 Design curves

The section capacity ratio defined in Equations (5.14) and (5.24) can be plotted against the CHS diameter-to-thickness ratio for a certain type of strengthening scheme (i.e., a certain number of longitudinal and transverse layers), proportioning factor (ξ), and fibre reinforcement factor (α). Some examples are shown in Figures 5.4 and 5.5.

Figure 5.4 shows the influence of steel yield stress on the strength curves for CFRP-reinforced steel CHS columns with strengthening scheme 2T2L, proportioning factor $\xi = 0.3$, and fibre reinforcement factor $\alpha = 0.3$. Figure 5.5 illustrates the influence of the fibre reinforcement factor for CFRP-strengthened CHS columns with strengthening scheme 2T2L,

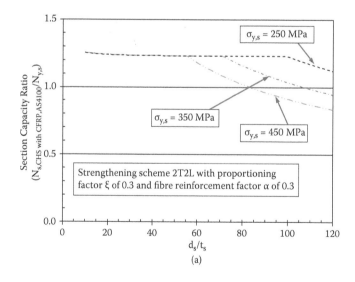

Figure 5.4 Examples of strength curves for CFRP-strengthened steel CHS columns with varying steel yield stress (e.g., strengthening scheme 2T2L). (a) Modified AS 4100 model. (Adapted from Haedir, J., and Zhao, X.L., *Journal of Constructional Steel Research*, 67(3), 497–509, 2011.)

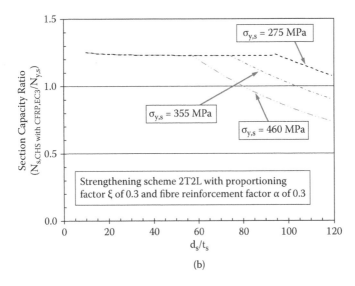

(b)

Figure 5.4 (Continued) Examples of strength curves for CFRP-strengthened steel CHS columns with varying steel yield stress (e.g., strengthening scheme 2T2L). (b) Modified EC3 model. (Adapted from Haedir, J., and Zhao, X.L., *Journal of Constructional Steel Research*, 67(3), 497–509, 2011.)

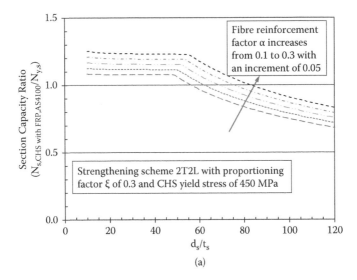

(a)

Figure 5.5 Examples of strength curves of AS 4100 and EC3-Part 1.1 for CFRP-strengthened steel CHS columns. (a) Modified AS 4100 model. (Adapted from Haedir, J., and Zhao, X.L., *Journal of Constructional Steel Research*, 67(3), 497–509, 2011.)

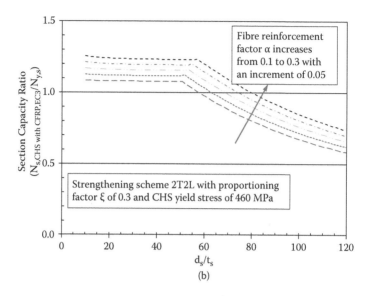

Figure 5.5 (Continued) Examples of strength curves of AS 4100 and EC3-Part 1.1 for CFRP-strengthened steel CHS columns. (b) Modified EC3 model. (Adapted from Haedir, J., and Zhao, X.L., *Journal of Constructional Steel Research*, 67(3), 497–509, 2011.)

proportioning factor ξ = 0.3, and yield stress 450 MPa for AS 4100 and 460 MPa for EC3-Part 1.1.

5.4.2 GFRP-strengthened CHS sections

A finite element (FE) analysis was conducted by Teng and Hu (2007) on GFRP-strengthened CHS stub columns, considering geometric and material nonlinearities, geometric imperfection, and an overlapping zone with thicker CFRP. This overlapping zone was treated in two different ways: the additional thickness of the overlapping zone (150 mm) was (1) directly included in the finite element model or (2) smeared around the tube. The predicted load versus axial shortening curves and failure modes agreed very well with the experimental observations.

A section capacity formula is proposed in this chapter for GFRP-strengthened CHS stub columns, based on the very limited amount of tests reported in Teng and Hu (2007). It reads:

$$N_{s,CHS \text{ with GFRP}} = k_n A_s \sigma_{y,s} \tag{5.27}$$

$$k_n = 1.05 \quad \text{for } n_{GFRP} = 1 \tag{5.28a}$$

$$k_n = 1.10 \quad \text{for } 1 < n_{GFRP} \leq 3 \tag{5.28b}$$

where A_s and $\sigma_{y,s}$ are the CHS cross-sectional area and yield stress, and n_{GFRP} is the number of GFRP layers.

The formula is only valid for CHS stub columns with a diameter-to-thickness ratio not less than 39, E_{GFRP} not less than 76 GPa, and $\sigma_{y,s}$ not greater than 350 MPa. More research is needed to develop more general formulae for CHS strengthened with GFRP.

5.4.3 CFRP-strengthened SHS sections

5.4.3.1 Bambach et al. stub column model

Bambach et al. (2009a) developed a design method whereby the theoretical elastic buckling stress of the composite steel-CFRP section is used to determine the axial section capacity. The design model was calibrated against 45 test results of using a reliability analysis approach. It is called the Bambach et al. stub column model in this chapter. The section capacity is given by

$$N_{s,SHS \text{ with CFRP}} = \rho_c A_s \sigma_{y,s} \qquad (5.29)$$

where $\sigma_{y,s}$ is the SHS yield stress and A_s is the cross-sectional area given by

$$A_s \approx b^2 - (b - 2t_s)^2 \qquad (5.30a)$$

or

$$A_s = 4(b - 2r_{ext})t_s + \pi(r_{ext}^2 - r_{int}^2) \qquad (5.30b)$$

where B, r_{ext}, and r_{int} are the SHS overall width, external corner radius, and inner corner radius. The latter can be estimated as (ASI 1999)

$$r_{int} = r_{ext} - t_s \qquad (5.31a)$$

$$r_{ext} = 2.5t_s \quad \text{for } t_s \geq 3 \text{ mm} \qquad (5.31b)$$

$$r_{ext} = 2.0t_s \quad \text{for } t_s < 3 \text{ mm} \qquad (5.31c)$$

The reduction factor (ρ_c) is taken from the winter equation, considering the composite steel-CFRP section properties:

$$\rho_c = \frac{1 - \dfrac{0.22}{\lambda_c}}{\lambda_c} \qquad (5.32)$$

$$\lambda_c = \sqrt{\frac{\sigma_{y,s}}{\sigma_{cr,cs}}} \qquad (5.33)$$

The elastic buckling stress of the composite plate ($\sigma_{cr,cs}$) is given by

$$\sigma_{cr,cs} = \frac{k\pi^2}{t_{total} \cdot (b - 2t_s)^2} D_t \tag{5.34}$$

where k is the plate buckling coefficient, which can be taken as 4.0, and t_{total} is the thickness of the composite section, given by

$$t_{total} = t_s + n(t_{CFRP\_sheet} + t_{a\_layer}) \tag{5.35}$$

As for the value of the transformed flexural rigidity (D_t) of the perfectly bonded two-layered plate, with each layer isotropic, it is obtained from the solution of a problem taken from Pister and Dong (1959) and defined by

$$D_t = \frac{D_1 D_3 - D_2^2}{D_1} \tag{5.36}$$

$$D_1 = \frac{E_s}{1 - v_s^2} t_s + \frac{E_{CFRP}}{1 - v_{CFRP}^2} (t_{total} - t_s) \tag{5.37a}$$

$$D_2 = \frac{E_s}{1 - v_s^2} \frac{t_s^2}{2} + \frac{E_{CFRP}}{1 - v_{CFRP}^2} \frac{t_{total}^2 - t_s^2}{2} \tag{5.37b}$$

$$D_3 = \frac{E_s}{1 - v_s^2} \frac{t_s^3}{3} + \frac{E_{CFRP}}{1 - v_{CFRP}^2} \frac{t_{total}^3 - t_s^3}{3} \tag{5.37c}$$

$$E_{CFRP} = \frac{E_a t_{a\_layer} + E_{CFRP\_sheet} t_{CFRP\_sheet}}{t_{a\_layer} + t_{CFRP\_sheet}} \tag{5.38}$$

where (1) E_{CFRP}, E_{CFRP_sheet}, and E_a are the Young's moduli of the CFRP composites, CFRP fibre, and adhesive; (2) t_{CFRP}, t_{CFRP_sheet}, and t_{a_layer} are the thicknesses of the CFRP composites, one layer of CFRP sheet, and one layer of adhesive; and (3) n is the number of CFRP layers.

It should be noted that the model described above applies a reduction factor to the whole cross-sectional area A_s. The reduction factor due to local buckling is normally not applied to the SHS corners. There are two ways to deal with the flat portion of SHS: to use either (1) the clear width ($b - 2t_s$) or (2) the flat width ($b - 2r_{ext}$). Then, Equation (5.29) can be modified as follows:

1. If the clear width concept is used,

$$N_{s,SHS \text{ with CFRP}} = 4\rho_c(b - 2t_s)t_s\sigma_{y,s} + 4t_s^2\sigma_{y,c} \tag{5.39}$$

where $\sigma_{y,c}$ is the corner yield stress, which can be estimated for cold-formed SHS as $1.22\sigma_{y,s}$ (Zhao et al. 2005b).

2. If the flat width concept is used,

$$N_{s,\text{SHS with CFRP}} = 4\rho_c(b - 2r_{ext})t_s\sigma_{y,s} + \pi(r_{ext}^2 - r_{int}^2)\sigma_{y,c} \qquad (5.40)$$

It should be noted that the reduction factor ρ_c cannot exceed 1.0. Therefore, the predicted section capacity will be limited to the yield capacity of the steel SHS. As mentioned in Bambach et al. (2009a), the CFRP is assumed to play an important role in the elastic buckling, but not in the strength.

The validity range of the parameters appearing in the Bambach et al. stub column model is summarised below:

$50 \leq b/t \leq 120$

$E_f \geq 230$ GPa

250 MPa $\leq \sigma_{y,s} \leq 450$ MPa

5.4.3.2 Shaat and Fam stub column model

Shaat and Fam (2006) proposed a simple formula to calculate the SHS stub column strengthened with FRP by adopting the cross-sectional area of the equivalent section (A_t). No local buckling reduction was considered because all the SHS studied had a width-to-thickness ratio below the limit specified in CAN/CSA 2001 (CSA 2001). The formula reads:

$$N_{s,\text{SHS with CFRP}} = A_t\sigma_{y,s} \qquad (5.41)$$

$$A_t = A_s + \frac{E_{CFRP}}{E_s}A_{CFRP} \qquad (5.42)$$

where A_s is the SHS cross-sectional area of steel, given in Equation (5.30), E_{CFRP} can be calculated using Equation (5.38), and A_{CFRP} is the area of the CFRP composites (i.e., combined fibres and adhesive), which can be calculated using different expressions:

1. If CFRP is applied on all four faces of SHS,

$$A_{CFRP} = 4 \times t_{CFRP} \times b_{CFRP} \qquad (5.43a)$$

2. If CFRP is applied on two faces of SHS,

$$A_{CFRP} = 2 \times t_{CFRP} \times b_{CFRP} \qquad (5.43b)$$

where t_{CFRP} is given by

$$t_{CFRP} = n(t_{CFRP\_sheet} + t_{a\_layer}) \qquad (5.44a)$$

$$b_{CFRP} = b - 2r_{ext} \qquad (5.44b)$$

in which t_{CFRP_sheet} is the thickness of one layer of CFRP sheet, t_{a_layer} is the thickness of one layer of adhesive, and n is the number of CFRP layers.

The validity range of the parameters appearing in the Shaat and Fam stub column model can be summarised as

$$14 \leq b/t_s \leq 28$$

$$\sigma_{y,s} \leq 380 \text{ MPa}$$

$$115 \text{ GPa} \leq E_f \leq 230 \text{ GPa}$$

$$kL/r < 5$$

5.4.4 CFRP-strengthened lipped channel sections

Silvestre et al. (2008, 2009) carried out nonlinear FE analyses of CFRP-strengthened lipped channel steel columns and developed design formulae to calculate their capacities by modifying the EC3-Part 1.3 (CEN 2006a) and AISI-DSM (2004) design rules. The proposed design rules are called the modified EC3 stub column model and modified AISI-DSM stub column model in this chapter.

5.4.4.1 Modified EC3 stub column model

The contribution of CFRP is considered by modifying the way to calculate the effective width of each compressed wall and by adding one capacity component concerning the CFRP. The section capacity is given by

$$N_{s,channel \text{ with CFRP}} = A_{eff,s}\sigma_{y,s} + A_{CFRP}\sigma_{CFRP} \qquad (5.45)$$

where $\sigma_{y,s}$ is the channel section yield stress, $A_{eff,s}$ is effective area, A_{CFRP} is the cross-sectional area of the CFRP composites, and σ_{CFRP} is the ultimate tensile strength of CFRP composites. The value of $A_{eff,s}$ is obtained from

$$A_{eff,s} = (b_{eff,web} + 2b_{eff,flange} + 2b_{eff,lip}) \cdot t_s \qquad (5.46)$$

The effective width for each component/wall (web, flange, and lip) can be calculated using Equations (5.47a) to (5.47c) given below, and the CFRP properties are included in the calculation of the slenderness λ_p, by means of Equation (5.48):

$$b_{eff,web} = \begin{cases} b_{web} & \text{if} \quad \lambda_p \leq 0.673 \\ \left(\dfrac{\lambda_p - 0.22}{\lambda_p^2}\right)b_{web} & \text{if} \quad \lambda_p > 0.673 \end{cases} \tag{5.47a}$$

$$b_{eff,flange} = \begin{cases} b_{flange} & \text{if} \quad \lambda_p \leq 0.673 \\ \left(\dfrac{\lambda_p - 0.22}{\lambda_p^2}\right)b_{flange} & \text{if} \quad \lambda_p > 0.673 \end{cases} \tag{5.47b}$$

$$b_{eff,lip} = \begin{cases} b_{lip} & \text{if} \quad \lambda_p \leq 0.748 \\ \left(\dfrac{\lambda_p - 0.188}{\lambda_p^2}\right)b_{lip} & \text{if} \quad \lambda_p > 0.748 \end{cases} \tag{5.47c}$$

$$\lambda_p = \sqrt{\dfrac{\sigma_{y,s}t_s + \sigma_{CFRP}t_{CFRP}}{\sigma_L\,(t_s + t_{CFRP})}} \tag{5.48}$$

where t_{CFRP} is the thickness of the CFRP composites (i.e., combining fibres and adhesive), given in Equation (5.44), and σ_L is the wall local buckling stress, which can be calculated according to Annex A.1 of EC3-Part 1.5 (CEN 2006b).

The modified EC3 stub column model is applicable to lipped channels with a yield stress not greater than 550 MPa and CFRP with a Young's modulus not less than 230 GPa.

5.4.4.2 Modified AISI-DSM stub column model

The direct strength method (DSM) provides an elegant, efficient, and consistent approach to estimate the ultimate strength of cold-formed steel columns (Schafer 2008). The current AISI specification (AISI 2007) allows the use of the DSM method for the design of columns against local, distortional, global, and local-global interactive failures. The influence of CFRP is considered by adding one capacity component, concerning the CFRP, to the yield load of the steel lipped channel sections. This modified yield load is also used to calculate the section slenderness (λ_c). The application of the method is based on the expressions

$$N_{s,channel\ with\ CFRP} = N_y 0.658^{\lambda_c^2} \tag{5.49}$$

$$N_y = A_s \sigma_{y,s} + A_{CFRP}\sigma_{CFRP} \tag{5.50}$$

$$\lambda_c = \sqrt{\frac{N_y}{N_E}} \qquad (5.51)$$

where N_E is the critical elastic buckling load defined in clause 1.2.1.1 of AISI (2007).

The modified AISI-DSM stub column model is applicable to lipped channel sections with a yield stress not greater than 550 MPa and CFRP with a Young's modulus not less than 230 GPa.

5.4.5 CFRP-strengthened T-sections

Harries et al. (2009) investigated the behaviour of FRP-strengthened T-sections in compression, but no design formulae were provided. The increase in section capacity was found to vary between 4 and 14%. On the basis of the limited amount of test results reported by Harries et al. (2009), it is possible to propose in this chapter simple formulae for T-sections strengthened by FRP:

$$N_{s,T \text{ with GFRP}} = k_{GFRP} N_{s,T} \qquad (5.52a)$$

$$N_{s,T \text{ with CFRP}} = k_{CFRP} N_{s,T} \qquad (5.52b)$$

$$k_{GFRP} = 1.05 \qquad (5.53a)$$

$$k_{CFRP} = 1.10 \qquad (5.53b)$$

where $N_{s,T}$ is the section capacity of a bare steel T-section (no FRP strengthening).

The above formulae are only valid for T-section stub columns with (1) a web depth-to-thickness ratio not less than 28 and (2) a GFRP Young's modulus above 41 GPa or CFRP Young's modulus above 235 GPa. More research is needed to enable the development of more general formulae for FRP-strengthened T-section columns.

5.5 CAPACITY OF CFRP-STRENGTHENED STEEL MEMBERS

5.5.1 CFRP-strengthened SHS columns

5.5.1.1 Fibre model and FE analysis

Shaat and Fam (2007a) developed a nonlinear fibre model to predict the load versus (axial and lateral) displacement curves of CFRP-strengthened SHS columns. The model takes into account material and geometric nonlinearities, residual stresses, and failure of CFRP in compression.

An element-by-element approach is adopted to integrate stresses over the steel and FRP cross-sectional areas. Due to symmetry, only one-half of the section needs to be analysed. Figure 5.6 shows a schematic view of a meshing system where A1 to A4 are the four areas considered. The flat part of the flange (A1) is divided into 12 strips through the thickness, whereas the flat part of the web (A2) is divided into 12 × 80 elements, to capture the strain gradient and residual stress distribution. The corner is idealised as a square (A3) and divided into 12 × 12 elements. Finally, area A4 represents the FRP layers, and each layer is one element.

The calculation requires an iterative process. A predicted load-axial displacement response is compared with experimental data in Figure 5.7. It is important to include the residual stress effects and the FRP failure in compression in the fibre model.

Shaat and Fam (2007b) carried out a nonlinear finite element analysis of CFRP-strengthened SHS columns. Similarly to the fibre model, the FE analysis includes geometric and material nonlinearities, residual stress distribution, and initial geometrical imperfection. A parametric study was also conducted to examine the influence of the slenderness ratios, imperfections, number of CFRP layers, and level of residual stresses on the column capacity. The column capacity improvement was found to be highly dependent on the column slenderness and CFRP reinforcement ratio ρ_f, defined in Equation (5.54d), rather than on the imperfection amplitude.

5.5.1.2 Shaat and Fam column model

Shaat and Fam (2009) developed a design-oriented model for CFRP-strengthened SHS columns, which consists of a modification of the

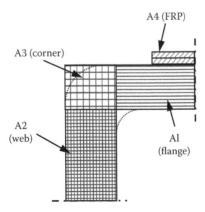

Figure 5.6 Schematic view of meshing system for CFRP-strengthened SHS columns. (Adapted from Shaat, A., and Fam, A., *Journal of Structural Engineering*, 133(1), 85–95, 2007a.)

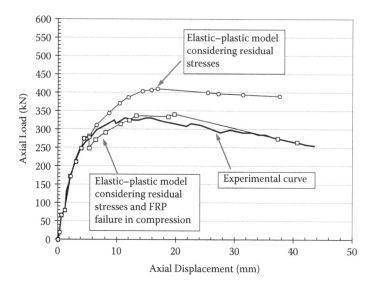

Figure 5.7 Illustration of theoretical models. (Adapted from Shaat, A., and Fam, A., *Journal of Structural Engineering*, 133(1), 85–95, 2007a.)

design rules for SHS columns included in the ANSI/AISC specification (AISC 2005). It is called the Shaat and Fam column model in this chapter.

The member capacity depends on the slenderness of the composite columns ($kL/r_t \rho_f$), where k is the column effective length factor, which depends on the end support conditions. The cross-sectional radius of gyration r_t and area ratio ρ_f can be calculated from:

$$r_t = \sqrt{\frac{I_t}{A_t}} \tag{5.54a}$$

$$A_t = A_s + \frac{E_{CFRP}}{E_s} A_{CFRP} \tag{5.54b}$$

$$I_t = I_s + \frac{E_{CFRP}}{E_s} I_{CFRP} \tag{5.54c}$$

$$\rho_f = \frac{A_{CFRP}}{A_s} \tag{5.54d}$$

where I_s is the SHS second moment of area and I_f is the second moment of area of CFRP composites, both about the axis parallel to the Y-Y axis defined in Figure 5.1(b) that passes through the cross section centre.

By comparing the average strain of the CFRP ($\varepsilon_{CFRP,c}$) with the average flexural buckling strain (ε_{cr}), Shaat and Fam (2009) were able to identify three possible failure modes: (1) CFRP fails before overall buckling, (2) CFRP fails simultaneously with overall buckling, and (3) CFRP fails after overall buckling. Empirical formulae were derived for the average strain of CFRP ($\varepsilon_{CFRP,c}$), which can be rewritten as

$$\frac{\varepsilon_{CFRP,c}}{\varepsilon_{u,CFRP}} = 0.00183\left(\frac{kL}{r_t\rho_f}\right) \quad \text{for } kL/r_t\rho_f \leq 224 \tag{5.55a}$$

$$\frac{\varepsilon_{CFRP,c}}{\varepsilon_{u,CFRP}} = -0.00054\left(\frac{kL}{r_t\rho_f}\right)+0.53 \quad \text{for } 224 \leq kL/r_t\rho_f \leq 315 \tag{5.55b}$$

$$\frac{\varepsilon_{CFRP,c}}{\varepsilon_{u,CFRP}} = -0.00146\left(\frac{kL}{r_t\rho_f}\right)+0.82 \quad \text{for } kL/r_t\rho_f \leq 315 \tag{5.55c}$$

where $\varepsilon_{u,CFRP}$ is the ultimate tensile strain of the CFRP composites.

The average flexural buckling strain (ε_{cr}) can be determined using clause E3 of AISC (2005), using the radius of gyration (r_t) of the composite section. No local buckling reduction is considered because all the SHSs studied had a width-to-thickness ratio lower than the limit specified in clause E7.2 of AISC (2005):

1. For $\dfrac{kL}{r_t} \leq 4.71\sqrt{\dfrac{E_s}{\sigma_{y,s}}}$

$$\varepsilon_{cr} = \left(0.658^{\sigma_{y,s}/\sigma_e}\right)\frac{\sigma_{y,s}}{E_s} \tag{5.56a}$$

2. For $\dfrac{kL}{r_t} > 4.71\sqrt{\dfrac{E_s}{\sigma_{y,s}}}$

$$\varepsilon_{cr} = \frac{0.877\sigma_e}{E_s} \tag{5.56b}$$

The elastic critical buckling stress (σ_e) is given by

$$\sigma_e = \frac{\pi^2 E_s}{\left(\dfrac{kL}{r_t}\right)} \tag{5.57}$$

where k is the column effective length factor, which depends on the end support conditions, L is the column length, and r_t is the radius of gyration of the composite section, given in Equation (5.54).

Three design cases are given below:

Case 1: If $\varepsilon_{CFRP,c} < \varepsilon_{cr}$,

$$N_{c,SHS\ with\ CFRP} = Max\{A_s E_s \varepsilon_{cr},\ A_s E_s \varepsilon_{CFRP,c} + A_{CFRP} E_{CFRP} \varepsilon_{CFRP,c}\} \quad (5.58a)$$

Case 2: If $\varepsilon_{CFRP,c} = \varepsilon_{cr}$,

$$N_{c,SHS\ with\ CFRP} = A_s E_s \varepsilon_{cr} + A_{CFRP} E_{CFRP} \varepsilon_{CFRP,c} \quad\quad\quad (5.58b)$$

Case 3: If $\varepsilon_{CFRP,c} > \varepsilon_{cr}$,

$$N_{c,SHS\ with\ CFRP} = A_s E_s \varepsilon_{cr} + A_{CFRP} E_{CFRP} \varepsilon_{cr} \quad\quad\quad (5.58c)$$

where A_s and A_{CFRP} are given in Equations (5.30) and (5.43), E_{CFRP} is given in Equation (5.38), and $\varepsilon_{CFRP,c}$ and ε_{cr} are given in Equations (5.55) and (5.56), respectively.

The validity range of the parameters appearing in the Shaat and Fam column model can be summarised as

$$14 \leq b/t_s \leq 28$$

$$\sigma_{y,s} \leq 380\ MPa$$

$$E_f \geq 230\ GPa$$

$$46 \leq kL/r \leq 93$$

$$kL/r_t \rho_f \leq 400$$

(only longitudinal CFRP is applied)

More research is needed to develop formulae for parameters falling outside of this range.

5.5.2 CFRP-strengthened lipped channel columns

Silvestre et al. (2009) developed design formulae for CFRP-strengthened lipped channel steel columns by modifying the EC3-Part 1.3 (CEN 2006a) and AISI-DSM (2004) design rules. The proposed design approaches are called the modified EC3 column model and modified AISI-DSM column model in Section 5.5.2.

5.5.2.1 Modified EC3 column model

The column capacity ($N_{c,channel\ with\ CFRP}$) is a product of the strength reduction factor and the section capacity described in Section 5.4.4.1, except that the influence of distortional buckling on the lip and flange thickness needs to be considered. The capacity is given by

$$N_{c,channel\ with\ CFRP} = \chi(A_{eff,s}\sigma_{y,s} + A_{CFRP}\sigma_{CFRP}) \tag{5.59}$$

$$A_{eff,s} = b_{eff,web}\ t_s + 2b_{eff,flange}t_{flange,red} + 2b_{eff,lip}t_{lip,red} \tag{5.60}$$

where χ is the strength reduction factor associated with global (flexural or flexural–torsional) buckling and based on buckling curve b (imperfection factor $\alpha_{LT} = 0.34$), as prescribed in EC3-Part 1.3 (CEN 2006a) for singly symmetric sections (buckling curve b is plotted in Figure 5.8 to assist the readers). The nondimensional slenderness λ_T is defined by

$$\lambda_T = \sqrt{\frac{A_{eff,s}\sigma_{y,s}}{N_{cr}}} \tag{5.61}$$

where $A_{eff,s}$ is given in Equation (5.60) and N_{cr} is the column elastic critical load associated with torsional or flexural–torsional buckling, given in clause 6.3.1.4 of EC3-Part 1.1 (CEN 2005).

The effective widths $b_{eff,web}$, $b_{eff,flange}$, and $b_{eff,lip}$ are given in Equations (5.47a) to (5.47c). A thickness reduction of the cross section "stiffeners" (lips and adjacent effective flange portions, i.e., $t_{flange,red}$ and $t_{lip,red}$ in Equation (5.60)) is adopted in EC3-Part 1.3 (CEN 2006a) to take into account the influence of distortional buckling. The reduced thickness can be calculated as follows:

$$t_{red} = t_s \quad \text{if } \lambda_d \leq 0.65 \tag{5.62a}$$

$$t_{red} = (1.47 - 0.723\lambda_d)t_s \quad \text{if } 0.65 < \lambda_d < 1.38 \tag{5.62b}$$

$$t_{red} = \left(\frac{0.66}{\lambda_d}\right)t_s \quad \text{if } \lambda_d \geq 1.38 \tag{5.62c}$$

$$\lambda_d = \sqrt{\frac{\sigma_{y,s}t_s + \sigma_{CFRP}t_{CFRP}}{\sigma_D\ (t_s + t_{CFRP})}} \tag{5.63}$$

where σ_D is the distortional buckling stress. It can be determined by either means of numerical methods, such as the shell finite element method, the finite strip method, or generalised beam theory, or using the simplified methodologies included in clause 5.5.3.2 of EC3-Part 1.3 (CEN 2006a). The determination of the distortional buckling stress is outside the scope of this book. Readers can find more information on distortional buckling, for instance, in Hancock (2007).

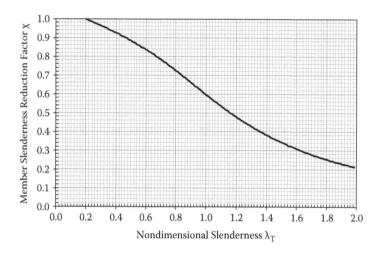

Figure 5.8 Buckling curve b from Eurocode 3 Part 1.3. (From CEN, *Design of Steel Structures—Eurocode 3, Part 1–3, General rules—Supplementary Rules for Cold-Formed Members and Sheeting*, EN 1993-1-3, European Committee for Standardization, Brussels, 2006).

5.5.2.2 Modified AISI-DSM column model

For CFRP-strengthened lipped channel columns subject to compressive forces, three possible failure modes exist: global, interactive local-global, and distortional failure. The column capacity is the least of the capacities corresponding to the three failure modes.

$$N_{c,\text{channel with CFRP}} = \text{Min}\{N_{c,E}, N_{c,L}, N_{c,D}\} \tag{5.64}$$

where $N_{c,E}$, $N_{c,L}$, and $N_{c,D}$ are the column nominal strengths against the elastic global, interactive local-global, and distortional failure, respectively. They are obtained from

$$N_{c,E} = \begin{cases} N_y 0.658^{\lambda_c^2} & \text{if} \quad \lambda_c \leq 1.5 \\[2mm] N_y \left(\dfrac{0.877}{\lambda_c^2} \right) & \text{if} \quad \lambda_c > 1.5 \end{cases} \tag{5.65}$$

where N_y is given in Equation (5.50) and λ_c is defined in Equation (5.51):

$$N_{c,L} = \begin{cases} N_{c,E} & \text{if } \lambda_L \leq 0.776 \\[2mm] N_{c,E} \left(\dfrac{N_L}{N_{c,E}} \right)^{0.4} \left[1 - 0.15 \left(\dfrac{N_L}{N_{c,E}} \right)^{0.4} \right] & \text{if } \lambda_L > 0.776 \end{cases} \tag{5.66a}$$

$$\lambda_L = \sqrt{\frac{N_{c,E}}{N_L}} \tag{5.66b}$$

where N_L is the column elastic critical local buckling load, defined in clause 1.2.1.2 of AISI (2007), and

$$N_{c,D} = \begin{cases} N_y & \text{if } \lambda_D \leq 0.561 \\[2ex] N_y \left(\frac{N_D}{N_y}\right)^{0.6} \left[1-0.25\left(\frac{N_D}{N_y}\right)^{0.6}\right] & \text{if } \lambda_D > 0.561 \end{cases} \tag{5.67a}$$

$$\lambda_D = \sqrt{\frac{N_y}{N_D}} \tag{5.67b}$$

where N_D is the column elastic critical distortional buckling load, defined in clause 1.2.1.3 of AISI (2007).

Appendix 1 of AISI (2007) allows the design of cold-formed sections using the DSM approach. The values of the critical loads N_E, N_L, and N_D can be obtained either analytically or numerically. The detailed description of DSM is outside the scope of this book and can be found in Schafer (2008) and in the commentary to the AISI specification (AISI 2007).

The modified EC3 and AISI-DSM column models are applicable to lipped channel sections with a yield stress not larger than 550 MPa and CFRP with a Young's modulus not lower than 230 GPa. Silvestre et al. (2009) found that the above equations provide good predictions of the capacity of CFRP-strengthened lipped channel columns if the composite layer thickness does not exceed 10 to 15% of the steel column wall thickness.

5.6 PLASTIC MECHANISM ANALYSIS OF CFRP-STRENGTHENED SHS UNDER LARGE AXIAL DEFORMATION

Plastic mechanism analysis has been widely used to study steel members and connections that involve local collapse mechanisms (Zhao 2003). It has been identified by Rasmussen and Hancock (1998) as one of the four fundamental methods of collapse analysis. This method has been used in the past to analyse SHS columns under large deformation (Kecman 1983, Key and Hancock 1986, Mahendran and Murray 1990, Grzebieta and White 1994, Kotelko et al. 2000, Zhao et al. 2002).

Bambach and Elchalakani (2007) applied this method to CFRP-strengthened SHS columns undergoing large deformation. The derivations are similar to those reported in the literature for SHS members in compression by Key and Hancock (1986) and Zhao et al. (2002). The influence of CFRP is considered through modifications of the material properties that are summarised in this section.

5.6.1 Equivalent yield stress due to CFRP strengthening

The equilibrium condition based on the stress distribution across the thickness, shown in Figure 5.9, implies that

$$\sigma_{y,s}(t_{total} - y_{na}) = \sigma_{y,s}(y_{na} - t_{CFRP}) + \sigma_{CFRP}t_{CFRP} \quad (5.68)$$

where y_{na} is the neutral axis position, which can be determined by

$$y_{na} = \frac{\sigma_{y,s}(t_{total} + t_{CFRP}) - \sigma_{CFRP}t_{CFRP}}{2\sigma_{y,s}} \quad (5.69)$$

The plastic moment capacity per unit width can be calculated by adding the moment generated from each force component:

$$
\begin{aligned}
M_p = \sigma_{y,s}\frac{(t_{total} - y_{na})^2}{2} + \sigma_{y,s}\frac{(y_{na} - t_{CFRP})^2}{2} \\
+ \sigma_{CFRP}t_{CFRP}\left(y_{na} - \frac{t_{CFRP}}{2}\right)
\end{aligned}
\quad (5.70)
$$

Equation (5.70) can be rewritten in terms of the dimensionless ratios k_σ and k_t, defined below (in Equations (5.72a) and (5.72b)) as

$$M_p = \left(1 + 2k_\sigma k_t + 2k_\sigma k_t^2 - k_\sigma^2 k_t^2\right)\frac{\sigma_{y,s}t_s^2}{4} \quad (5.71)$$

$$k_\sigma = \frac{\sigma_{CFRP}}{\sigma_{y,s}} \quad (5.72a)$$

$$k_t = \frac{t_{CFRP}}{t_s} \quad (5.72b)$$

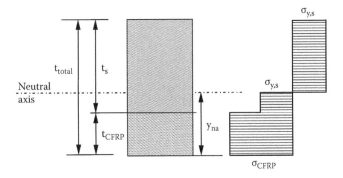

Figure 5.9 Stress distribution. (Adapted from Bambach, M.R., and Elchalakani, M., *Thin-Walled Structures*, 45(2), 159–170, 2007.)

The equivalent yield stress for the face $(\sigma_{y,s,eq})$ can be derived using the relationship

$$M_p = \frac{\sigma_{y,s,eq}t_{total}^2}{4} = \left(1 + 2k_\sigma k_t + 2k_\sigma k_t^2 - k_\sigma^2 k_t^2\right)\frac{\sigma_{y,s}t_s^2}{4} \qquad (5.73)$$

$$\sigma_{y,s,eq} = \frac{1 + 2k_\sigma k_t + 2k_\sigma k_t^2 - k_\sigma^2 k_t^2}{\left(1 + k_t\right)^2}\sigma_{y,s} \qquad (5.74)$$

Similarly, the equivalent yield stress for the corner $(\sigma_{y,c,eq})$ can be derived as

$$\sigma_{y,c,eq} = \frac{1 + 2k_{\sigma,c}k_t + 2k_{\sigma,c}k_t^2 - k_{\sigma,c}^2 k_t^2}{\left(1 + k_t\right)^2}\sigma_{y,c} \qquad (5.75a)$$

$$k_{\sigma,c} = \frac{\sigma_{CFRP}}{\sigma_{y,c}} \qquad (5.75b)$$

where $\sigma_{y,c}$ is the corner yield stress, which can be estimated for cold-formed SHS as $1.22\sigma_{y,s}$ (Zhao et al. 2005b).

5.6.2 Plastic mechanism analysis

The plastic mechanism assumed by Bambach and Elchalakani (2007) is illustrated in Figure 5.10. The total load carrying capacity $(N_{SHS\ with\ CFRP})$ of the stub column is a sum of all the load components $(P_A, P_B, and P_C)$ shown in Figure 5.10 and carried by the stub column:

$$N_{SHS\ with\ CFRP} = 4P_A + 4P_B + 4P_C \qquad (5.76)$$

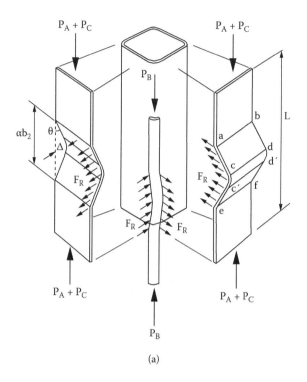

(a)

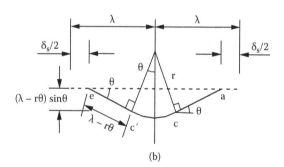

(b)

Figure 5.10 Load components and axial deformation in plastic mechanism analysis.
(a) Load components. (b) Deformation. (From Bambach, M.R., and Elchalakani,
M., *Thin-Walled Structures*, 45(2), 159–170, 2007.)

From the face folding mechanism shown in Figure 5.10, the reduced moment capacity, due to axial loading (P_A), of a plastic hinge can be written as

$$M_p' = \frac{\sigma_{y,s,eq}b_2t_s^2}{4}\left(1-\left(\frac{P_A}{\sigma_{y,s,eq}b_2t_s}\right)^2\right) \tag{5.77}$$

A more general discussion on reduced moment capacity of plastic hinges can be found in Zhao and Hancock (1993).

From the equilibrium condition

$$P_A = \frac{2M_p'}{(\lambda\sin\theta - r\theta\sin\theta)} \tag{5.78}$$

it can be concluded that

$$P_A = \sigma_{y,s,eq}b_2t_s\left\{\sqrt{\left(\frac{\lambda\sin\theta - r\theta\sin\theta}{t_s}\right)^2 + 1} - \left(\frac{\lambda\sin\theta - r\theta\sin\theta}{t_s}\right)\right\} \tag{5.79}$$

where the symbols (b_2, λ, θ, r) are defined in Figure 5.10:

$$\lambda \approx \frac{b_2}{2} \tag{5.80a}$$

$$b_2 = b - 2r_{ext} \tag{5.80b}$$

$$r \approx 3t_s \tag{5.80c}$$

where b is the overall SHS flange width and r_{ext} is the external corner radius defined in Equation (5.31).

The corner yielding load (P_B) can be simply expressed as

$$P_B = \sigma_{y,c}A_{corner} = \sigma_{y,c}t_s\frac{\pi}{2}\frac{r_{ext} + r_{int}}{2} = \frac{\sigma_{y,c}t_s(r_{ext} + r_{int})}{4} \tag{5.81}$$

where r_{ext} and r_{int} are the cold-formed SHS external and internal corner radii, defined in Equation (5.31). The contribution of the CFRP to the compression yielding in the axial direction is assumed to be small, and hence was ignored in Equation (5.81).

The load component P_C can be derived using the virtual work principle (details given in Bambach and Elchalakani 2007), which leads to

$$P_C = \frac{\sigma_{y,c,eq}t_s\lambda^2}{10(\lambda\sin\theta - r\theta\sin\theta)} \tag{5.82}$$

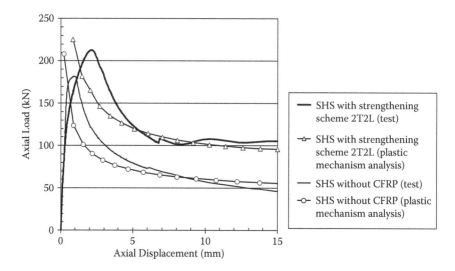

Figure 5.11 Comparison between predicted and experimental post-collapse curves. (Adapted from Bambach, M.R., and Elchalakani, M., *Thin-Walled Structures*, 45(2), 159–170, 2007.)

where $\sigma_{y,c,eq}$ is given in Equation (5.75), and λ and r are given in Equations (5.80a) and (5.80c).

Figure 5.11 displays predicted and experimental post-collapse curves of SHS columns, which show a reasonable agreement.

5.7 DESIGN EXAMPLES

5.7.1 Example 1: CFRP-strengthened CHS stub column

Determine the section capacity of a cold-formed C350 circular hollow section (CHS 101.6 × 2.6) strengthened using scheme 2T2L defined in Section 5.2. The CFRP sheet has a thickness of 0.176 mm and Young's moduli of 230 GPa (longitudinal direction) and 14 GPa (transverse direction). The CHS has a yield stress of 350 MPa and Young's modulus of 200 GPa.

5.7.1.1 Solution using the modified AS 4100 model given in Section 5.4.1.1

1. Dimension and properties:
   $d_s = 101.6$ mm
   $t_s = 2.6$ mm

$E_s = 200$ GPa

$T_{CFRP\_sheet} = 0.176$ mm

$E_{L,t,cs} = 230$ GPa

$E_{T,t,cs} = 14$ GPa

$n_L = 2$

$n_T = 2$

$\sigma_{y,s} = 350$ MPa

$$A_s = \frac{1}{4} \cdot \pi \cdot d_s^2 - \frac{1}{4} \cdot \pi \cdot (d_s - 2 \cdot t_s)^2$$

$$= \frac{1}{4} \times 3.1416 \times 101.6^2 - \frac{1}{4} \times 3.1416 \times (101.6 - 2 \times 2.6)^2$$

$$= 809\,\text{mm}^2$$

2. Check validity range:

$d_s/t_s = 101.6/2.6 = 39.0$, which is between 37 and 78—OK

$\sigma_{y,s} = 350$ MPa, which is between 250 and 450 MPa—OK

$\alpha = 0.27$ (as shown below), which is between 0.1 and 0.3—OK

All conditions are satisfied.

3. Proportioning factor and fibre reinforcement factor:

Using Equation (5.8),

$$\xi = \frac{E_{T,t,cs}}{E_{L,t,cs}} = \frac{14}{230} = 0.061$$

Using Equation (5.11),

$$\alpha = \frac{(n_L + n_T)t_{CFRP\_sheet}}{t_s} = \frac{(2+2) \cdot 0.176}{2.6} = 0.27$$

4. Diameter-to-thickness ratio of the equivalent section:

From Equation (5.6),

$$\beta_L = \frac{E_{L,t,cs}}{E_s} = \frac{230}{200} = 1.15$$

The additional thickness due to CFRP strengthening is determined using Equation (5.10):

$t_{es,cs} = \beta_L t_{CFRP\_sheet}(n_L + \xi n_T) = 1.15 \times 0.176 \times (2 + 0.061 \times 2) = 0.429$ mm

The thickness of the equivalent section becomes

$t_{es} = t_s + t_{es,cs} = 2.6 + 0.429 = 3.03$ mm

The diameter of the equivalent section becomes

$d_{es} = d_s + 2t_{es,cs} = 101.6 + 2 \times 3.03 = 107.66$ mm

Hence,

$$\frac{d_{es}}{t_{es}} = \frac{107.66}{3.03} = 35.5$$

5. Section capacity:
   The limit of diameter-to-thickness ratio is

$$\frac{82}{\left(\dfrac{\sigma_{y,s}}{250}\right)} = \frac{82}{\left(\dfrac{350}{250}\right)} = 58.6$$

Since

$$\frac{d_{es}}{t_{es}} = 35.5 < \frac{82}{\left(\dfrac{\sigma_{y,s}}{250}\right)} = 58.6$$

Equation (5.16a) should be used:

$$N_{s,CHS \text{ with } CFRP,AS4100} \approx \left(1 + \frac{t_{es,cs}}{t_s}\right) \cdot A_s \sigma_{y,s} = \left(1 + \frac{0.429}{2.6}\right) \times 809 \times 350$$

$$= 329{,}870\,N \approx 330\,kN$$

5.7.1.2 Solution using the modified EC3 model given in Section 5.4.1.2

1. Dimension and properties:
   Same as those given in Section 5.7.1.1.
2. Check validity range:
   All conditions are satisfied as shown in Section 5.7.1.1.
3. Proportioning factor and fibre reinforcement factor:
   Same as those in given Section 5.7.1.1.

   $\xi = 0.061$
   $\alpha = 0.27$
4. Diameter-to-thickness ratio of the equivalent section:
   Same as those given in Section 5.7.1.1.

$$\frac{d_{es}}{t_{es}} = 35.5$$

5. Section capacity:
   The limit of diameter-to-thickness ratio is

$$\frac{90}{\left(\dfrac{\sigma_{y,s}}{235}\right)} = \frac{90}{\left(\dfrac{350}{235}\right)} = 60.4$$

Since

$$\frac{d_{es}}{t_{es}} = 35.5 < \frac{90}{\left(\dfrac{\sigma_{y,s}}{235}\right)} = 60.4$$

Equation (5.26a) should be used:

$$N_{s,CHS \text{ with CFRP,EC3}} \approx \left(1 + \frac{t_{es,cs}}{t_s}\right) \cdot A_s\sigma_{y,s} = \left(1 + \frac{0.429}{2.6}\right) \times 809 \times 350$$

$$= 329,870\,N \approx 330\,kN$$

It can be seen that the modified EC3 model gives the same section capacity as the modified AS 4100 model. This is because the equivalent CHS section in this example has a diameter-to-thickness ratio less than the local buckling limit in both codes, which implies that Equations (5.16a) and (5.26a) are identical.

5.7.2 Example 2: CFRP-strengthened SHS stub column with local buckling

Determine the section capacity of a cold-formed square hollow section (SHS $100 \times 100 \times 2$) strengthened using scheme 2T2L defined in Figure 5.1(a). The CFRP sheet has a Young's modulus of 230 GPa and a thickness of 0.176 mm. The adhesive has a Young's modulus of 1.9 GPa and a thickness of 0.1 mm per layer. The SHS has a yield stress of 400 MPa and a Young's modulus of 200 GPa. Poisson's ratio is 0.3 for steel and 0.25 for CFRP. The ultimate tensile strain of CFRP composites is assumed to be 1.2%.

Here is the solution using the Bambach et al. stub column model given in Section 5.4.3.1.

1. Dimension and properties:
    $b = 100$ mm
    $t_s = 2$ mm
    $E_s = 200$ GPa
    $\upsilon_s = 0.3$
    $t_{CFRP\_sheet} = 0.176$ mm
    $E_{CFRP\_sheet} = 230$ GPa
    $\upsilon_{CFRP} = 0.25$
    $t_{a\_layer} = 0.1$ mm
    $E_a = 1.9$ GPa
    $n = 2 + 2 = 4$
    $\sigma_{y,s} = 400$ MPa
    $A_s \approx b^2 - (b - 2t_s)^2 = 100^2(100 - 2\times 2)^2 = 784$ mm^2

2. Check validity range:
   $b/t_s = 100/2 = 50$, which is the same as the lower limit of 50—OK
   $\sigma_{y,s} = 400$ MPa, which is between 250 and 450 MPa—OK
   $E_f = 230$ GPa, which is the same as the lower limit of 230 GPa—OK
   $\xi = 0.3$ (as shown below), which is less than 0.8—OK
   $\alpha = 0.27$ (as shown below), which is between 0.1 and 0.3—OK
   All conditions are satisfied.
3. Properties of the composite section:
   Using Equation (5.38),

$$E_{CFRP} = \frac{E_a t_{a\_layer} + E_{CFRP\_sheet} t_{CFRP\_sheet}}{t_{a\_layer} + t_{CFRP\_sheet}} = \frac{1.9 \times 0.1 + 230 \times 0.176}{0.1 + 0.176}$$

$$= 147\,GPa$$

From Equation (5.35),

$$t_{total} = t_s + n(t_{CFRP\_sheet} + t_{a\_layer}) = 2 + 4 \times (0.176 + 0.1) = 3.104 \text{ mm}$$

4. Elastic buckling stress of the composite plate:
   From Equation (5.37),

$$D_1 = \frac{E_s}{1-v_s^2} t_s + \frac{E_{CFRP}}{1-v_{CFRP}^2}(t_{total} - t_s) = \frac{200}{1-0.3^2} \times 2$$

$$+ \frac{230}{1-0.25^2} \times (3.104 - 2) = 710$$

$$D_2 = \frac{E_s}{1-v_s^2} \frac{t_s^2}{2} + \frac{E_{CFRP}}{1-v_{CFRP}^2} \frac{t_{total}^2 - t_s^2}{2} = \frac{200}{1-0.3^2} \times \frac{2^2}{2}$$

$$+ \frac{230}{1-0.25^2} \times \frac{3.104^2 - 2^2}{2} = 1131$$

$$D_3 = \frac{E_s}{1-v_s^2} \frac{t_s^3}{3} + \frac{E_{CFRP}}{1-v_{CFRP}^2} \frac{t_{total}^3 - t_s^3}{3} = \frac{200}{1-0.3^2} \times \frac{2^3}{3}$$

$$+ \frac{230}{1-0.25^2} \times \frac{3.104^3 - 2^3}{3} = 2378$$

Using Equation (5.36),

$$D_t = \frac{D_1 D_3 - D_2^2}{D_1} = \frac{710 \times 2378 - 1131^2}{710} = 576$$

From Equation (5.34),

$$\sigma_{cr,cs} = \frac{k\pi^2}{t_{total} \cdot (b - 2t_s)^2} D_t = \frac{4 \times 3.1416^2}{3.104 \times (100 - 2 \times 2)^2} \times 576$$

$$= 0.795 \text{ GPa} = 795 \text{ MPa}$$

5. Reduction factor (ρ_c):
From Equation (5.33),

$$\lambda_c = \sqrt{\frac{\sigma_{y,s}}{\sigma_{cr,cs}}} = \sqrt{\frac{400}{795}} = 0.709$$

Using Equation (5.32),

$$\rho_c = \frac{1 - \dfrac{0.22}{\lambda_c}}{\lambda_c} = \frac{1 - \dfrac{0.22}{0.709}}{0.709} = 0.973$$

6. Section capacity:
From Equation (5.29),

$$N_{s,SHS \text{ with CFRP}} = \rho_c A_s \sigma_{y,s} = 0.973 \times 784 \times 400 = 30513 \text{ N} \approx 305 \text{ kN}$$

If the clear width concept is used, from Equation (5.39),

$$N_{s,SHS \text{ with CFRP}} = 4\rho_c (b - 2t_s) t_s \sigma_{y,s} + 4t_s^2 \sigma_{y,c}$$

$$= 4 \times 0.973 \times (100 - 2 \times 2) \times 2 \times 400 + 4 \times 2^2 \times (1.22 \times 400)$$

$$= 306714 \text{ N}$$

$$= 307 \text{ N}$$

If the flat width concept is used, from Equation (5.31),
$r_{ext} = 2.0t_s = 2.0 \times 2 = 4$ mm
$r_{int} = r_{ext} - t_s = 4 - 2 = 2$ mm
From Equation (5.40),

$$N_{s,SHS \text{ with CFRP}} = 4\rho_c (b - 2r_{ext}) t_s \sigma_{y,s} + \pi (r_{ext}^2 - r_{int}^2) \sigma_{y,c}$$

$$= 4 \times 0.973 \times (100 - 2 \times 4) \times 2 \times 400$$

$$+ 3.1416 \times (4^2 - 2^2) \times (1.22 \times 400)$$

$$= 304848 \text{ N}$$

$$= 305 \text{ N}$$

It seems that there is very little difference between the section capacities calculated using the three approaches given in Equations (5.29), (5.39), and (5.40).

5.7.3 Example 3: CFRP-strengthened SHS stub column without local buckling

Determine the section capacity of a cold-formed square hollow section (SHS 50 × 50 × 3) strengthened with two layers of longitudinal CFRP sheets on two opposite faces. The CFRP sheet has a Young's modulus of 230 GPa and a thickness of 0.176 mm. The adhesive has a Young's modulus of 1.9 GPa and a thickness of 0.1 mm for each layer. The SHS has a yield stress of 350 MPa and Young's modulus of 200 GPa. The length of the SHS stub column is 190 mm. Both ends of the SHS stub columns are fixed, which corresponds to an effective length factor k = 0.5.

Here is the solution using the Shaat and Fam stub column model given in Section 5.4.3.2.

1. Dimension and properties:

   b = 50 mm

   t_s = 3 mm

   E_s = 200 Gpa

   t_{CFRP_sheet} = 0.176 mm

   E_{CFRP_sheet} = 230 Gpa

   t_{a_layer} = 0.1 mm

   E_a = 1.9 Gpa

   n = 2

   L = 190 mm

   $\sigma_{y,s}$ = 350 MPa

   k = 0.5

   r_{ext} = 2.5t_s = 2.5 × 3 = 7.5 mm

   $A_s \approx b^2 - (b - 2t_s)^2 = 50^2 - (50 - 2 \times 3)^2 = 564$ mm^2

   $$I_s = \frac{b^4}{12} - \frac{(b-2t_s)^4}{12} = \frac{50^4}{12} - \frac{(50-2\times3)^4}{12} = 0.209 \times 10^6 \text{ mm}^4$$

   $$r_s = \sqrt{\frac{I_s}{A_s}} = \sqrt{\frac{0.209 \times 10^6}{564}} = 19.2 \text{ mm}$$

2. Check validity range:

   b/t = 50/3 = 16.7, which is between 14 and 28—OK

   $\sigma_{y,s}$ = 350 MPa ≤ 380 MPa—OK

   E_f = 230 GPa ≤ 230 GPa—OK

   kL/r = 0.5 × 190/19.2 = 4.95 < 5—OK

   All conditions are satisfied.

3. Properties of the composite section:

   Using Equation (5.38),

$$E_{CFRP} = \frac{E_a t_{a\_layer} + E_{CFRP\_sheet} t_{CFRP\_sheet}}{t_{a\_sheet} + t_{CFRP\_sheet}} = \frac{1.9 \times 0.1 + 230 \times 0.176}{0.1 + 0.176}$$

$$= 147 \text{ GPa}$$

From Equation (5.44),

$t_{CFRP} = n(t_{CFRP\_sheet} + t_{a\_layer}) = 2 \times (0.176 + 0.1) = 0.552$ mm

$b_{CFRP} = b - 2r_{ext} = 50 - 2 \times 7.5 = 35$ mm

From Equation (5.43b),
$A_{CFRP} = 2 \times t_{CFRP} \times b_{CFRP} = 2 \times 0.552 \times 35 = 38.64$ mm^2
From Equation (5.42),

$$A_t = A_s + \frac{E_{CFRP}}{E_s} A_{CFRP} = 564 + \frac{147}{200} \times 38.64 \approx 592 \text{ mm}^2$$

4. Section capacity:
   Hence the section capacity can be determined using Equation (5.41),

   $N_{s,\text{SHS with CFRP}} = A_t \sigma_{y,s} = 592 \times 350 = 207200 \text{ N} \approx 207 \text{ kN}$

5.7.4 Example 4: CFRP-strengthened SHS slender column

Determine the member capacity of a cold-formed square hollow section (SHS 50 × 50 × 3) strengthened with four layers of longitudinal CFRP sheets on two opposite faces. The CFRP sheet has a Young's modulus of 230 GPa and a thickness of 0.176 mm. The adhesive has a Young's modulus of 1.9 GPa and a thickness of 0.1 mm per layer. The SHS has a yield stress of 350 MPa and a Young's modulus of 200 GPa. The length of the SHS column is 1250 mm. The column is fixed at one end and pinned at the other end with an effective length factor k approximated as 0.8.

Here is the solution using the Shaat and Fam column model given in Section 5.5.1.2.

1. Dimension and properties:
   b = 50 mm
   t_s = 3 mm
   E_s = 200 Gpa
   t_{CFRP_sheet} = 0.176 mm
   E_{CFRP_sheet} = 230 Gpa
   t_{a_layer} = 0.1 mm
   E_a = 1.9 Gpa
   n = 4
   L = 1250 mm

$\sigma_{y,s}$ = 350 MPa

k = 0.8

r_{ext} = 2.5t_s = 2.5 × 3 = 7.5 mm

$A_s \approx b^2 - (b - 2t_s)^2 = 50^2 - (50 - 2 \times 3)^2 = 564$ mm²

$$I_s = \frac{b^4}{12} - \frac{(b-2t_s)^4}{12} = \frac{50^4}{12} - \frac{(50-2\times3)^4}{12} = 0.209\times10^6 \text{ mm}^4$$

$$r_s = \sqrt{\frac{I_s}{A_s}} = \sqrt{\frac{0.209\times10^6}{564}} = 19.2 \text{ mm}$$

2. Check validity range:

   b/t = 50/3 = 16.7, which is between 14 and 28—OK

   $\sigma_{y,s}$ = 350 MPa ≤ 380 MPa—OK

   E_f = 230 GPa ≥ 230 GPa—OK

   kL/r_s = 0.8 × 1250/19.2 = 52.1, which is between 46 and 93—OK

   kL/$r_t\rho_f$ = 290 (shown below in step 5) < 400—OK

   Only longitudinal CFRP is applied—OK

   All conditions are satisfied.

3. Properties of the composite section:

   Using Equation (5.38),

   $$E_{CFRP} = \frac{E_a t_{a\_layer} + E_{CFRP\_sheet} t_{CFRP\_sheet}}{t_{a\_layer} + t_{CFRP\_sheet}} = \frac{1.9\times0.1 + 230\times0.176}{0.1 + 0.176}$$

   $$= 147 \text{ GPa}$$

   From Equation (5.44),

   $$t_{CFRP} = n(t_{CFRP\_sheet} + t_{a\_layer}) = 4 \times (0.176 + 0.1) = 1.104 \text{ mm}$$

   $$b_{CFRP} = b - 2r_{ext} = 50 - 2 \times 7.5 = 35 \text{ mm}$$

   From Equation (5.43b),

   $$A_{CFRP} = 2 \times t_{CFRP} \times b_{CFRP} = 2 \times 1.104 \times 35 = 77.28 \text{ mm}^2$$

   The second moment of area of CFRP (I_{CFRP}) is the sum of the contributions from each face. It is calculated about the axis passing through the centre of the CFPR component and parallel to Y-Y axis defined in Figure 5.1(b). Then,

   $$I_{CFRP} \approx 2 \times \frac{b_{CFRP} t_{CFRP}^3}{12} + 2 \times b_{CFRP} t_{CFRP} \cdot \left(\frac{b}{2} + \frac{t_{CFRP}}{2}\right)^2$$

   $$= 2 \times \frac{35\times1.104^3}{12} + 2\times35\times1.014\times\left(\frac{50}{2} + \frac{1.104}{2}\right)^2$$

   $$= 0.051\times10^6 \text{ mm}^4$$

   From Equation (5.54),

$$A_t = A_s + \frac{E_{CFRP}}{E_s} A_{CFRP} = 564 + \frac{147}{200} \times 77.28 = 620.8 \text{ mm}^2$$

$$I_t = I_s + \frac{E_{CFRP}}{E_s} I_{CFRP} = 0.209 \times 10^6 + \frac{147}{200} \times 0.051 \times 10^6$$

$$= 0.247 \times 10^6 \text{ mm}^4$$

$$r_t = \sqrt{\frac{I_t}{A_t}} = \sqrt{\frac{0.247 \times 10^6}{620.8}} = 19.95 \text{ mm}$$

$$\rho_f = \frac{A_{CFRP}}{A_s} = \frac{77.28}{564} = 0.137$$

4. Elastic critical buckling stress:
   The elastic critical buckling (average) stress is determined using Equation (5.57). For simply supported columns, the effective length factor k is equal to 1.0 and σ_e reads

$$\sigma_e = \frac{\pi^2 E_s}{\left(\dfrac{kL}{r_t}\right)} = \frac{3.1416^2 \times 200,000}{\left(\dfrac{0.8 \times 1250}{19.95}\right)} = 39,380 \text{ MPa}$$

5. Strains:
   The average flexural buckling strain (ε_{cr}) is determined by Equation (5.56) and depends on the value of kL/r_t.
   It is necessary to check the conditions in Equation (5.56) because

$$\frac{kL}{r_t} = \frac{0.8 \times 1250}{19.95} = 50.1 < 4.71\sqrt{\frac{E_s}{\sigma_{y,s}}} = 4.71\sqrt{\frac{200,000}{350}} = 113$$

From Equation (5.56a)

$$\varepsilon_{cr} = \left(0.658^{\sigma_{y,s}/\sigma_e}\right)\frac{\sigma_{y,s}}{E_s} = \left(0.658^{350/39,380}\right) \times \frac{350}{200,000} = 1.74 \times 10^{-3}$$

It is necessary to check the conditions in Equation (5.55) because

$$\left(\frac{kL}{r_t \rho_f}\right) = \left(\frac{0.8 \times 1250}{19.95 \times 0.137}\right) \approx 366$$

which exceeds 315.
   Equation (5.55c) should be used:

$$\varepsilon_{CFRP,c} = \left(-0.00146 \left(\frac{kL}{r_t \rho_f} \right) + 0.82 \right) \varepsilon_{u,CFRP}$$

$$= \left(-0.00146 \times \left(\frac{0.8 \times 1250}{19.95 \times 0.137} \right) + 0.82 \right) \times 0.012$$

$$= 3.43 \times 10^{-3}$$

6. Member capacity:
   It can be found that case 3, i.e., $\varepsilon_{CFRP,c} > \varepsilon_{cr}$, holds, which means that Equation (5.58c) should be used:

$$N_{c,SHS \text{ with CFRP}} = A_s E_s \varepsilon_{cr} + A_{CFRP} E_{CFRP} \varepsilon_{cr}$$

$$= 564 \times 200,000 \times 1.74 \times 10^{-3}$$

$$+ 77.28 \times 230,000 \times 1.74 \times 10^{-3}$$

$$= 227,199 \text{ N}$$

$$= 227 \text{ kN}$$

5.8 FUTURE WORK

There is a need to conduct more experimental tests and numerical simulation to cover wider ranges of parameters, such as column cross section shapes and dimensions, member slenderness, CFRP moduli, adhesive types, and thickness. More research is also needed to understand the behaviour of FRP-strengthened metallic columns subject to impact and blast loading (Bambach et al. 2009b), as well as to harsh environmental conditions (Seica and Packer 2007). Another important area of research is about connections between CFRP-strengthened compression members and beams or connections between CFRP-strengthened compression braces and chord members in trusses.

More recently, some researchers began to explore the possibility of repair/strengthening of concrete-filled steel tubes (CFSTs) using CFRP. Tests were carried out (Zhao et al. 2005a, Tao et al. 2007a, Sun et al. 2008) on concrete-filled steel hollow section short columns strengthened by CFRP. The dominating failure mode was found to be CFRP rupture at outward mechanism locations, which was also observed for CFRP-strengthened hollow section columns (see Figure 5.2(c)). The increase in load carrying capacity varies from 5 to 44% depending on the number of CFRP layers and section slenderness. Tao et al. (2007a) found that less increase in load carrying capacity due to CFRP strengthening was achieved for concrete-filled rectangular hollow sections, although a similar increase in ductility was found for both CHSs and RHSs. Similar work was reported in Hu et al. (2011) on GFRP-strengthened concrete-filled circular hollow sections.

Xiao et al. (2005) studied circular CFST columns confined by CFRP to avoid plastic hinges forming at critical locations in buildings (i.e., soft storey behaviour) under earthquake loading. A gap (made with 1 mm thick soft foam tapes affixed on the surface of the steel tube) was introduced between the CFRP and the tubular column to delay the engagement of the CFRP, to achieve both increased strength and ductility. The static strength increased by 55 and 140% when the number of CFRP layers was two and four, respectively. The ductility with the gap was found to be twice that without the gap. Seismic behaviour of CFST columns can be significantly improved by providing additional confinement to the potential hinge region. The local buckling and subsequent rupture of the steel tube were effectively delayed compared with the counterpart CFST specimens. Similar work has been done by Park et al. (2010) on CFRP-confined square CFST columns. The confined column capacity increased up to 12%, while the ductility increased from 11 to 107%. For confined beam-columns an increase up to 20% was observed for both load carrying capacity and ductility.

A series of tests were conducted (Tao et al. 2007b, 2008, Tao and Han 2007) to investigate the feasibility of using CFRP in repairing CFST columns after exposure to fire. The strength enhancement from CFRP confinement decreased with increasing of eccentricity or the slenderness ratio. At the same time, the influence of CFRP repair on the stiffness was not apparent due to the fact that the confinement from CFRP wraps was moderate when the CFST beam-columns remained in an elastic stage. To some extent, ductility enhancement was observed, except those axially loaded shorter specimens with rupture of CFRP jackets at the mid-height occurred near peak loads. Improvement was also found in the cyclic performance of fire-damaged CFST columns repaired with CFRP. Tao et al. (2011) carried out fire endurance tests on FRP-strengthened circular CFST columns. The longitudinal fibres of the CFRP in an eccentrically loaded column could participate in resisting the bending moment more effectively. An insulation coating over CFRP could effectively delay the failure of composite columns.

Teng et al. (2007) developed a hybrid FRP-concrete-steel double-skin tubular column. The test results confirmed that the concrete in the new column is very effectively confined by the two tubes, and the local buckling of the inner steel tube is either delayed or suppressed by the surrounding concrete, leading to a very ductile response. When such hybrid sections are used as beams, the inner steel tube could be shifted toward the tensile side. Such beams were found to have a very ductile behaviour (Yu et al. 2006). The FRP tube enhances the structural behaviour by providing confinement to the concrete and additional shear resistance. Han et al. (2010) performed cyclic tests on such hybrid double-skin columns. It was found that such columns exhibit high levels of energy dissipation prior to the rupture of the longitudinal FRP.

More tests are needed on FRP-CFST columns and FRP-concrete-HSS hybrid columns with wider ranges of parameters. There is a need to develop analytical models, numerical simulation, and a design guide for such hybrid sections (Choi and Xiao 2010, Yu et al. 2010). Research is necessary to understand the behaviour of such hybrid sections subject to dynamic loading (Shan et al. 2007) and environmental conditions (Li et al. 2012).

REFERENCES

AISC. 2005. *Specification for structural steel buildings.* ANSI/AISC 360-05. Chicago: American Institute of Steel Construction.

AISI. 2004. *North American specification (NAS) for the design of cold-formed steel structural members: Design of cold-formed steel structural members with the direct strength method.* Appendix I. Washington, DC: American Iron and Steel Institute.

AISI. 2007. *North American specification for the design of cold-formed steel structural members.* Washington, DC: American Iron and Steel Institute.

ASI. 1999. *Design capacity tables for structural steel: Hollow sections.* Vol. 2. Sydney: Australian Steel Institute.

Bambach, M.R., and Elchalakani, M. 2007. Plastic mechanism analysis of steel SHS strengthened with CFRP under large axial deformation. *Thin-Walled Structures*, 45(2), 159–170.

Bambach, M.R., Elchlakani, M., and Zhao, X.L. 2009b. Composite steel–CFRP SHS tubes under axial impact. *Composite Structures*, 87(3), 282–292.

Bambach, M.R., Jama, H.H., and Elchalakani, M. 2009a. Axial capacity and design of thin-walled steel SHS strengthened with CFRP. *Thin-Walled Structures*, 47(10), 1112–1121.

CEN. 2005. *Design of steel structures. Eurocode 3, Part 1-1. General rules and rules for buildings.* EN 1993-1-1. Brussels: European Committee for Standardization.

CEN. 2006a. *Design of steel structures. Eurocode 3, Part 1-3. General rules—Supplementary rules for cold-formed members and sheeting.* EN 1993-1-3. Brussels: European Committee for Standardization.

CEN. 2006b. *Design of steel structures. Eurocode 3, Part 1-5. Plated structural elements.* EN 1993-1-5. Brussels: European Committee for Standardization.

Choi, K., and Xiao, Y. 2010. Analytical model of circular CFRP confined concrete-filled steel tubular columns under axial compression. *Journal of Composites for Construction*, 14(1), 125–133.

CSA. 2001. Limit states design of steel structures. Standard CAN/CSA S16-01. Mississauga, Ontario: Canadian Standards Association.

Grzebieta, R.H., and White, G.J. 1994. Void filled square cantilever steel tubes subject to gross plastic deformation. In *Proceedings of 6th International Symposium on Tubular Structures*, Melbourne, pp. 255–262.

Haedir, J., and Zhao, X.L. 2011. Design of short CFRP-reinforced steel tubular columns. *Journal of Constructional Steel Research*, 67(3), 497–509.

Han, L.H., Tao, Z., Liao, F., and Xu, Y. 2010. Tests on cyclic performance of FRP–concrete–steel double-skin tubular columns. *Thin-Walled Structures*, 48(6), 430–439.

Hancock, G.J. 2007. *Design of cold-formed steel structures*. 4th ed. Sydney: Australian Steel Institute.

Harries, K.A., Peck A.J., and Abraham, E.J. 2009. Enhancing stability of structural steel sections using FRP. *Thin-Walled Structures*, 47(10), 1092–1101.

Hu, Y.M., Yu, T., and Teng, J.G. 2011. FRP-confined circular concrete-filled thin steel tubes under axial compression. *Journal of Composites for Construction*, 15(5), 850–860.

Kecman, D. 1983. Bending collapse of rectangular and square section tubes. *International Journal of Mechanical Sciences*, 25(9–10), 623–636.

Key, P.W., and Hancock, G.J. 1986. Plastic collapse mechanisms for cold-formed square hollow section columns. In *Proceedings of the 10th Australasian Conference on the Mechanics of Structures and Materials*, Adelaide, pp. 217–222.

Kotelko, M., Lim, T.H., and Rhodes, J. 2000. Post-failure behaviour of box section beams under pure bending (an experimental study). *Thin-Walled Structures*, 38(2), 179–194.

Li, H., Ma, M., Xian, G., Yan, X., and Ou, J. 2012. Performances of concrete-filled GFRP or GFRP-steel circular tubes subjected to freeze-thaw cycles. *International Journal of Structural Stability and Dynamics*, 12(1), 95–108.

Mahendran, M., and Murray, N.W. 1990. Ultimate load behaviour of box columns under combined loading of axial compression and torsion. *Thin-Walled Structures*, 9(1–4), 91–120.

Park, J.W., Hong, Y.K., and Choi, S.M. 2010. Behaviors of concrete filled square steel tubes confined by carbon fiber sheets (CFS) under compression and cyclic loads. *Steel and Composite Structures*, 10(2), 187–205.

Pister, K.S., and Dong, S.B. 1959. Elastic bending of layered plates. *Journal of the Engineering Mechanics Division*, 85(4), 1–10.

Rasmussen, K.J., and Hancock, G.J. 1998. Buckling analysis of thin-walled structures: Analytical developments and applications. *Progress in Structural Engineering and Materials*, 1(3), 316–322.

Schafer, B.W. 2008. Review: The direct strength method of cold-formed steel member design. *Journal of Constructional Steel Research*, 64(7–8), 766–778.

Seica, M.V., and Packer, J.A. 2007. FRP materials for the rehabilitation of tubular steel structures, for underwater applications. *Composite Structures*, 80(3), 440–450.

Shaat, A. 2007. Structural behaviour of steel columns and steel–concrete composite girders retrofitted using CFRP. PhD thesis, Department of Civil, Construction and Environmental Engineering, North Carolina State University, Raleigh.

Shaat, A., and Fam, A. 2006. Axial loading tests on CFRP-retrofitted short and long HSS steel columns. *Canadian Journal of Civil Engineering*, 33(4), 458–470.

Shaat, A., and Fam, A. 2007a. Fiber-element model for slender HSS columns retrofitted with bonded high-modulus composites. *Journal of Structural Engineering*, 133(1), 85–95.

Shaat, A., and Fam, A. 2007b. Finite element analysis of slender HSS columns strengthened with high modulus composites. *Steel and Composite Structures*, 7(1), 19–34.

Shaat, A., and Fam, A. 2009. Slender steel columns strengthened using high-modulus CFRP plates for buckling control. *Journal of Composites for Construction*, 13(1), 2–12.

Shan, J.H., Chen, R., Zhang, W.X., Xiao, Y., Yi, W.J., and Lu, F.Y. 2007. Behavior of concrete filled tubes and confined concrete filled tubes under high speed impact. *Advances in Structural Engineering—An International Journal*, 10(2), 209–218.

Silvestre, N., Camotim, D., and Young, B. 2009. On the use of the EC3 and AISI specifications to estimate the ultimate load of CFRP-strengthened cold-formed steel lipped channel columns. *Thin-Walled Structures*, 47(10), 1102–1111.

Silvestre, N., Young, B., and Camotim, D. 2008. Non-linear behaviour and load-carrying capacity of CFRP-strengthened lipped channel steel columns. *Engineering Structures*, 30(10), 2613–2630.

Standards Australia. 1998. *Steel structures*. Australian Standard AS 4100. Sydney: Standards Australia.

Sun, G.S., Zhao, Y.H., and Gu, W. 2008. Stability of concrete filled CFRP–steel tube under axial compression. In *Proceedings of the 12th International Symposium on Tubular Structures*, Shanghai, China, October, pp. 111–116.

Tao, Z., and Han, L.H. 2007. Behaviour of fire-exposed concrete-filled steel tubular beam-columns repaired with CFRP wraps. *Thin-Walled Structures*, 45(1), 63–76.

Tao, Z., Han, L.H., and Wang, L.L. 2007b. Compressive and flexural behaviour of CFRP repaired concrete-filled steel tubes after exposure to fire. *Journal of Constructional Steel Research*, 63(8), 1116–1126.

Tao, Z., Han, L.H., and Zhuang, J.P. 2007a. Axial loading behavior of CFRP strengthened concrete-filled steel tubular sub columns. *Advances in Structural Engineering—An International Journal*, 10(1), 37–46.

Tao, Z., Han, L.H., and Zhuang, J.P. 2008. Cyclic performance of fire-damaged concrete-filled steel tubular beam-columns repaired with CFRP wraps. *Journal of Constructional Steel Research*, 64(1), 37–50.

Tao, Z., Wang, Z.B., Han, L.H., and Uy, B. 2011. Fire performance of concrete-filled steel tubular columns strengthened by CFRP. *Steel and Composite Structures*, 11(4), 307–324.

Teng, J.G., and Hu, Y.M. 2007. Behaviour of FRP-jacketed circular steel tubes and cylindrical shells under axial compression. *Construction and Building Materials*, 21(4), 827–838.

Teng, J.G., Yu, T., Wong, Y.L., and Dong, S.L. 2007. Hybrid FRP–concrete–steel tubular columns: Concept and behaviour. *Construction and Building Materials*, 21(4), 846–854.

Xiao, Y., He, W.H., and Choi, K.K. 2005. Confined concrete-filled tubular columns. *Journal of Structural Engineering*, 131(3), 488–497.

Yu, T., Teng, J.G., and Wong, Y.L. 2010. Stress–strain behavior of concrete in hybrid FRP-concrete-steel double-skin tubular columns. *Journal of Structural Engineering*, 136(4), 379–389.

Yu, T., Wong, L., Teng, J.G., Dong, S., and Lam, E. 2006. Flexural behavior of hybrid FRP-concrete-steel double-skin tubular members. *Journal of Composites for Construction*, 10(5), 443–452.

Zhao, X.L. 2003. Yield line mechanism analysis of steel members and connections. *Progress in Structural Engineering and Materials*, 5(4), 252–262.

Zhao, Y.H., Gu, W., Xu, J., and Zhang, H.T. 2005a. The strength of concrete-filled CFRP–steel tubes under axial compression. In *Proceedings of the 15th International Offshore and Polar Engineering Conference*, Seoul, Paper JSC-313.

Zhao, X.L., Han, B., and Grzebieta, R.H. 2002. Plastic mechanism analysis of concrete-filled double-skin (SHS inner and SHS outer) stub columns. *Thin-Walled Structures*, 40(10), 815–833.

Zhao, X.L., and Hancock, G.J. 1993. Experimental verification of the theory of plastic moment capacity of an inclined yield line under axial force. *Thin-Walled Structures*, 15(3), 209–233.

Zhao, X.L., Wilkinson, T., and Hancock, G.J. 2005b. *Cold-formed tubular members and connections*. Oxford: Elsevier.

Zhao, X.L., and Zhang, L. 2007. State of the art review on FRP strengthened steel structures. *Engineering Structures*, 29(8), 1808–1823.

Chapter 6

Strengthening of web crippling of beams subject to end bearing forces

6.1 GENERAL

Web crippling of thin-walled steel members is often observed at loading or reaction points where concentrated forces exist (see examples shown in Figure 6.1). The types of steel members could be in the form of cold-formed rectangular hollow section (RHS), aluminium RHS, LiteSteel beam (LSB), channel, or I-sections, as shown in Figure 6.2.

Extensive research was carried out in the past on web crippling of cold-formed RHS, aluminium RHS, LSB, channel, and I-sections subjected to concentrated bearing forces (Packer 1984, 1987, Zhao and Hancock 1992, 1995, Wilkinson et al. 2006a, Zhou et al. 2009, Young and Hancock 2001, Young and Zhou 2008, Zhou and Young 2008, 2010). Web crippling consists of two failure modes: web buckling and web yielding. The key parameters governing the behaviour include the web depth-to-thickness ratio and corner radius. The flanges in different types of sections provide a certain amount of restraint against web rotation, which represents different boundary conditions for web buckling. The external corner radius (r_{ext}) in cold-formed RHS introduces load eccentricity to the webs, which reduces the web crippling capacity. Another important parameter influencing web crippling behaviour is the loading position, i.e., interior bearing where the distance between the edge of the section and the loading point is larger than 1.5d or otherwise end bearing. In general, the web crippling capacity of end bearing is lower than that for interior bearing. This chapter deals with the end bearing, which is the worst case.

Attempts were made in the past to increase the web crippling capacity of cold-formed RHS. The technique used includes (1) partially filling the RHS with wood plus a bolt through the web (Zhao 1999) and (2) partially filling the RHS with concrete (Packer and Fear 1991, Zhao 1999). Wilkinson et al. (2006b) developed several methods to increase web buckling capacity of LSB by attaching steel plates or inserting square hollow sections (SHSs) into the web or using proprietary brackets. In the case of I-section members, it is common to provide welded transverse stiffeners to prevent web

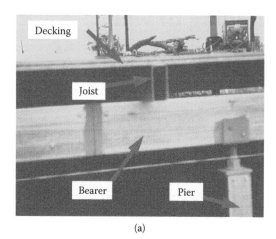

(a)

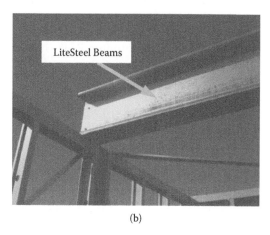

(b)

Figure 6.1 Examples of beams subjected to end bearing forces. (a) Cold-formed RHS in DuraGal floor system. (Courtesy of OneSteel, Australia.) (b) LiteSteel beams used in a domestic house. (Courtesy of Smorgon Steel Tube Mills, Australia.)

crippling. For cold-formed channel section purlins, the cleats that attach the members to support structures provide web stiffening (Hancock 1994, Young and Hancock 2001).

This chapter deals with the use of carbon fibre-reinforced polymer (CFRP) to increase web crippling capacity of cold-formed RHS, aluminium RHS, LSB, channel, and I-section beams. Three types of strengthening schemes are adopted: type O, where CFRP is applied outside the web; type I, where CFRP is applied inside the web; and type B, where CFRP is applied on both sides of the web. The strengthening of each section shown in Figure 6.2 will be described one by one in terms of types of strengthening, failure modes,

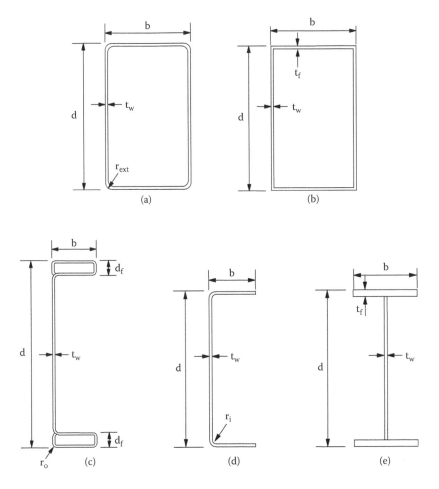

Figure 6.2 Section dimensions of beams to be strengthened: (a) cold-formed RHS, (b) aluminium RHS, (c) LiteSteel beams, (d) channel section, and (e) I-section.

load-displacement behaviour, increased capacity, and design formulae. Some design examples will also be given.

6.2 COLD-FORMED STEEL RECTANGULAR HOLLOW SECTIONS

6.2.1 Types of strengthening

Zhao et al. (2006) carried out a series of experimental testing on cold-formed RHSs strengthened by CFRP plates. A schematic view of three types of strengthening techniques is shown in Figure 6.3. One CFRP plate

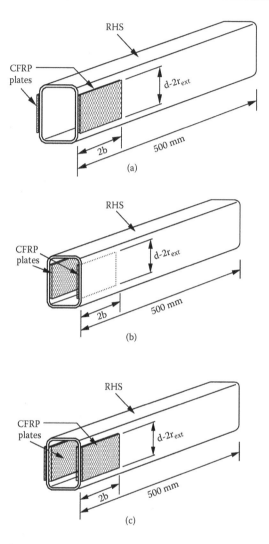

Figure 6.3 Type of strengthening for cold-formed RHS: (a) type O (outside web strengthening), (b) type I (inside web strengthening), and (c) type B (both sides strengthening). (From Zhao, X.L. et al., *Engineering Structures*, 28(11), 1555–1565, 2006.)

is bonded to each web outside of the RHS (called type O), or bonded to each web inside the RHS (called type I). Type B is a combination of type O and type I. Grinding was used to remove the galvanised coating and other impurities from the RHS surface. Then the RHS surface was cleaned using acetone before applying adhesive. Araldite A and Araldite B were mixed according to the recommended weight ratio given by the manufacturer. A thin coat was uniformly applied on the surface along the bond length. The CFRP was then placed and slowly rubbed in a circular motion to remove any air bubbles. The specimens were kept at room temperature for 1 week, followed by keeping them in an oven at 60°C for another week.

In all the strengthening the direction of CFRP fibre is always perpendicular to the longitudinal axis of RHS. A length of CFRP plates up to twice the overall width (b) of the RHS flange was found sufficient because the length of failure mechanism in the web observed in the previous tests (Zhao and Hancock 1995) on unstrengthened RHS is less than 2b.

6.2.2 Failure modes

A series of photos are presented in Figure 6.4 to show the progress of failure in a typical specimen for type B strengthening. It can be seen that CFRP plates are engaged in load carrying at an early stage. Web buckling is prevented by CFRP plates. In general, the delamination phenomenon commences at the first peak load. For type O strengthening the delamination developed relatively quicker than that for type I and type B, where inner CFRP plates exist.

Figure 6.5(a) shows the failure mode of unstrengthened RHS. For the first two sections the failure mode is web buckling due to a large web depth-to-thickness ratio. For the last section the failure mode is web yielding because of its small web depth-to-thickness ratio. The failure modes of these sections after strengthening using type O are shown in Figure 6.5(b). A section with d/t of 50 after type O strengthening still failed in web buckling, as shown in Figure 6.5(b)(i). Web yielding was achieved for sections with smaller d/t ratios after type O strengthening. Web yielding failure mode was observed for all sections after type I and type B strengthening, as shown in Figure 6.5(c) and (d).

6.2.3 Behaviour

Typical load versus deformation curves are plotted in Figure 6.6 for a cold-formed RHS 100 × 50 × 2 (i.e., depth of 100 mm, width of 50 mm, and thickness of 2 mm) with various types of strengthening. The deformation in Figure 6.6 is defined as the vertical movement of loading plate. It can be seen from Figure 6.6 that CFRP strengthening increases not only the ultimate load capacity, but also the ductility of RHS member. For type O and type I strengthening, there is a drop in load carrying capacity after the first peak, followed by an increase in load to reach the second peak. The ultimate load carrying capacity obtained for type I is higher than that for type O. This is mainly due to the fact that the inner CFRP plate reduces the load eccentricity transferred to the RHS web, especially when

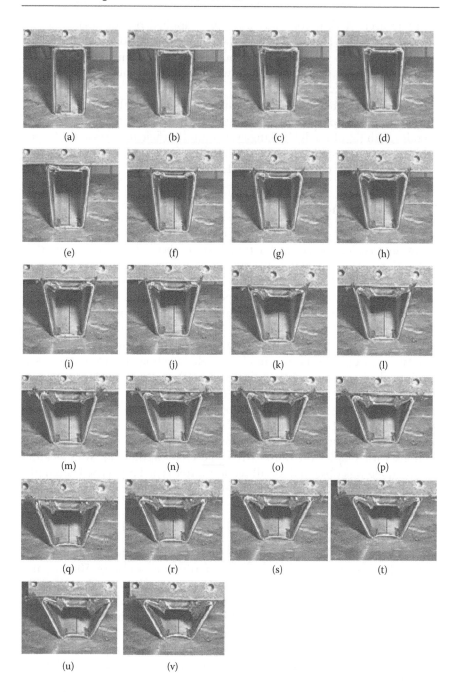

Figure 6.4 An example of progress of failure (cold-formed RHS with type B strengthening). (Courtesy of D. Fernando, Monash University, Australia.)

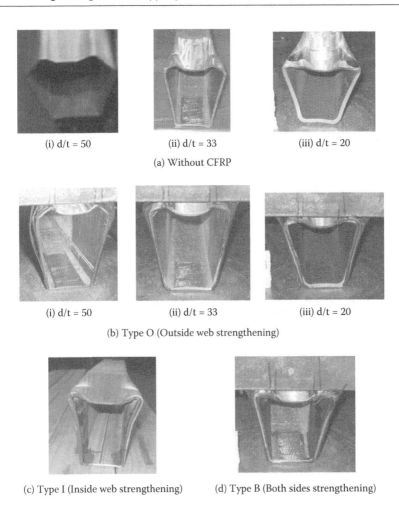

(i) d/t = 50   (ii) d/t = 33   (iii) d/t = 20

(a) Without CFRP

(i) d/t = 50   (ii) d/t = 33   (iii) d/t = 20

(b) Type O (Outside web strengthening)

(c) Type I (Inside web strengthening)   (d) Type B (Both sides strengthening)

Figure 6.5 Typical failure modes for each strengthening type (cold-formed RHS). (Courtesy of D. Fernando, Monash University, Australia.)

the inner CFRP starts to deform with the corners. The load drop after the first peak in type B is insignificant with an even higher ultimate load carrying capacity.

6.2.4 Increased capacity

The increase of bearing capacity due to three types of strengthening is plotted in Figure 6.7 against the overall depth-to-thickness ratio (d/t_w). It seems that type O produces about 50% increase no matter what the d/t_w ratio

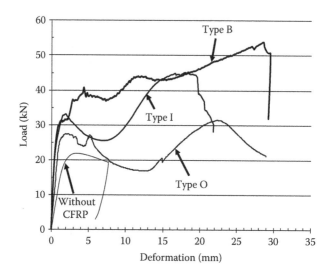

Figure 6.6 Typical load versus deformation curves (cold-formed RHS 100 × 50 × 2 with various types of strengthening). (Adapted from Zhao, X.L. et al., *Engineering Structures*, 28(11), 1555–1565, 2006.)

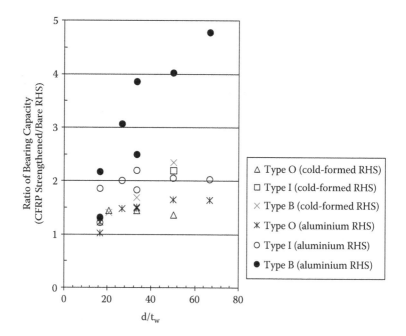

Figure 6.7 Increase in bearing capacity for cold-formed RHS and aluminium RHS.

is, whereas thinner sections benefit more from type B strengthening. This is most likely because of the change of failure mode from web buckling to web yielding in the case of type B strengthening.

6.2.5 Design formulae

6.2.5.1 Design formulae for unstrengthened RHS

Design formulae for CFRP-strengthened cold-formed RHS subject to end bearing forces have been proposed by Zhao et al. (2006). They are a modification of design rules for unstrengthened RHS given in the Australian Standard AS 4100 (Standards Australia 1998). For cold-formed RHS subject to end bearing forces, two failure modes need to be checked: web bearing buckling and web bearing yielding. The web bearing capacity (R_b) of a cold-formed RHS is the lesser of the web bearing buckling capacity (R_{bb}) and web bearing yield capacity (R_{by}), i.e.,

$$R_b = \min \{R_{bb}, R_{by}\} \tag{6.1}$$

The formulae for R_{bb} and R_{by} of unstrengthened RHS are summarised below for the convenience of readers. The dimensions are defined in Figure 6.8.

$$R_{bb} = 2 \cdot b_b \cdot t_w \cdot f_y \cdot \alpha_c \tag{6.2}$$

$$R_{by} = 2 \cdot b_b \cdot t_w \cdot f_y \cdot \alpha_p \tag{6.3}$$

$$b_b = b_s + 0.5 \cdot d + 1.5 \cdot r_{ext} \tag{6.4}$$

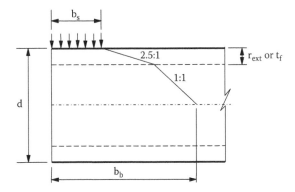

Figure 6.8 Load dispersion for RHS and I-section subject to end bearing force.

The external corner radius (r_{ext}) can be estimated (ASI 1999) as

$$r_{ext} = 2.5 \cdot t_w \qquad \text{for } t_w \geq 3 \text{ mm} \qquad (6.5a)$$

$$r_{ext} = 2.0 \cdot t_w \qquad \text{for } t_w < 3 \text{ mm} \qquad (6.5b)$$

α_c is the member slenderness reduction factor determined from AS 4100 for column buckling with section constant $\alpha_b = 0.5$ and a modified slenderness.

$$\lambda_n = k_e \cdot \sqrt{12} \cdot \left(\frac{d - 2 \cdot r_{ext}}{t_w} \right) \cdot \sqrt{\frac{f_y}{250}} \qquad (6.6)$$

$$\alpha_p = \sqrt{(2 + k_s^2)} - k_s \qquad (6.7a)$$

$$k_s = \frac{2 \cdot r_{ext}}{t_w} - 1 \qquad (6.7b)$$

where t_w is the web wall thickness, r_{ext} is the external corner radius, and f_y is the yield stress of the RHS. The graph of α_c is plotted in Figure 6.9 to assist readers. The effective buckling length factor k_e is taken as 1.1, as in AS 4100, for unstrengthened RHS subject to end bearing forces.

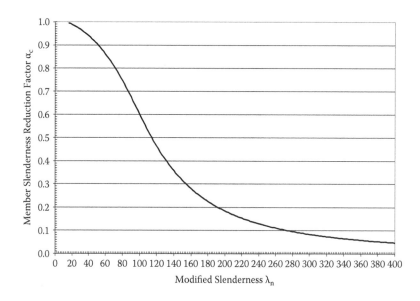

Figure 6.9 Member slenderness reduction factor for web buckling design (based on AS 4100). (From Standards Australia, *Steel Structures*, AS 4100, Australia, Sydney, 1998.)

The design of CFRP-strengthened RHS depends on the governing failure mode of unstrengthened RHS (web bearing buckling or web bearing yielding) and types of strengthening (type O, type I, or type B defined in Figure 6.3). Some guidance was given in Section 6.3.4 of Zhao et al. (2005) to determine whether an RHS design is governed by buckling or yielding by checking the following conditions:

For RHS with $t_w < 3$ mm:

$$\text{when } \left(\frac{d - 2 \cdot r_{ext}}{t_w}\right) \cdot \sqrt{\frac{f_y}{250}} > 36 \text{ web buckling governs} \tag{6.8a}$$

For RHS with $t_w \geq 3$ mm:

$$\text{when } \left(\frac{d - 2 \cdot r_{ext}}{t_w}\right) \cdot \sqrt{\frac{f_y}{250}} > 42 \text{ web buckling governs} \tag{6.8b}$$

6.2.5.2 Design formulae for CFRP-strengthened RHS (if web buckling governs for unstrengthened RHS)

Zhao et al. (2006) found that when a cold-formed RHS is governed by web buckling (i.e., satisfying Equation (6.8)), the CFRP-strengthened RHS can be designed as follows:

1. For type O strengthening (defined in Figure 6.3(a)): The failure mode after type O strengthening may remain as web buckling (see Figure 6.4 (b)(i)) or change to web yielding (see Figure 6.4(b)(ii)). Therefore, the design of such RHS is similar to that of unstiffened RHS, i.e., checking both web buckling and web yielding using Equations (6.1) to (6.7). The only difference is that a new effective buckling length factor k_e of 0.8 should be used. The reduced k_e value represents the enhanced restraining against web rotation.
2. For type I or type B strengthening (defined in Figure 6.3(b) and (c)): The failure mode after type I and type B strengthening changes to web yielding (see Figure 6.4(c)). Therefore, Equation (6.3) can be utilised to design such an RHS except that an α_p of 0.32 should be used. This adopted upper bound value of 0.32 represents the fact that CFRP plates provide restraints that minimise the eccentricity effect. The modified formulae can be rewritten as

$$R_{by} = 2 \cdot b_b \cdot t \cdot f_y \cdot \alpha_p \tag{6.9a}$$

$$b_b = b_s + 0.5 \cdot d + 1.5 \cdot r_{ext} \tag{6.9b}$$

$$\alpha_p = 0.32 \tag{6.9c}$$

6.2.5.3 Design formulae for CFRP-strengthened RHS
(if web yielding governs for unstrengthened RHS)

Zhao et al. (2006) found that when a cold-formed RHS is governed by web yielding (i.e., not satisfying Equation (6.8)), the CFRP-strengthened RHS can be designed as follows.

Because the failure mode remains web yielding after CFRP strengthening, an upper bound value of α_p (= 0.32) is adopted, as explained in Section 6.2.5.2. In addition, CFRP plates also create membrane action in steel plates that enter the strain-hardening stage. Hence, the ultimate tensile strength (f_u) may be used to replace the yield stress (f_y) in Equation (6.9). The modified formulae can be rewritten as

$$R_{by} = 2 \cdot b_b \cdot t \cdot f_u \cdot \alpha_p \tag{6.10a}$$

$$b_b = b_s + 0.5 \cdot d + 1.5 \cdot r_{ext} \tag{6.10b}$$

$$\alpha_p = 0.32 \tag{6.10c}$$

6.3 ALUMINIUM RECTANGULAR HOLLOW SECTIONS

6.3.1 Types of strengthening

An experimental program on using CFRP to strengthen aluminium RHS was conducted by Zhao and Phipatet (2009) and Wu et al. (2012). The strengthening scheme is very much the same as that described in Section 6.2 for cold-formed RHS (see Figure 6.3). Three types of strengthening methods (type O, type I, and type B) are shown in Figure 6.10.

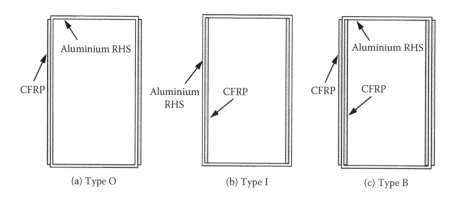

(a) Type O   (b) Type I   (c) Type B

Figure 6.10 Types of strengthening for aluminium RHS.

(a) Without CFRP   (b) Type O

(c) Type I   (d) Type B

Figure 6.11 Typical failure modes for each strengthening type (aluminium RHS). (Courtesy of P. Phipatet, Monash University, Australia.)

6.3.2 Failure modes

Typical failure modes for aluminium RHS without CFRP and with different types of strengthening are shown in Figure 6.11. Conventional web buckling was observed for aluminium RHS without CFRP, as shown in Figure 6.11(a). For type O and type I strengthening, an interesting observation is that all type O specimens buckled outward away from the section centre (see Figure 6.11(b)), while specimens with type I strengthening buckled inward toward the section centre (see Figure 6.11(c)). This is because of the load eccentricity generated by CFRP plates, which creates an extra bending moment. It can be seen that web buckling of aluminium RHS is effectively delayed by type O and type I strengthening, and prevented after type B strengthening. All type B specimens failed by web yielding, as shown in Figure 6.11(d).

6.3.3 Behaviour

Typical load versus deformation curves are presented in Figure 6.12 for aluminium RHS $100 \times 50 \times 3$ and $50 \times 50 \times 3$ with various CFRP strengthening.

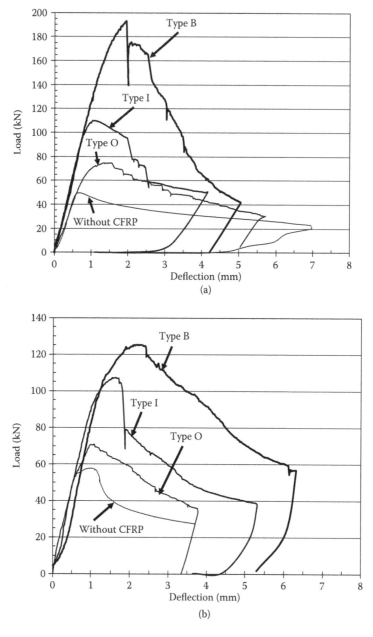

Figure 6.12 Typical load versus deformation curves (aluminium RHS with various types of strengthening). (a) Aluminium RHS 100 × 50 × 3. (b) Aluminium RHS 50 × 50 × 3. (Adapted from Wu, C. et al., *International Journal of Structural Stability and Dynamics*, 12(1), 109–130, 2012.)

It can be seen from Figure 6.12 that significant increases in load carrying capacity and ductility were obtained due to CFRP strengthening for all scenarios. Similar to cold-formed RHS strengthening, type I strengthening seems to be more effective than type O strengthening with a higher peak load. This is probably related to its failure mode, explained in Section 6.3.2.

6.3.4 Increased capacity

The bearing capacity of CFRP-strengthened aluminium RHS is compared in Figure 6.7 with that without CFRP. The increase due to type O and type I is about 50% and 100%, respectively. The increase due to type B ranges from 110 to 380%. More increase in bearing capacity is obtained for more slender (larger d/t_w) RHSs.

6.3.5 Design formulae

6.3.5.1 Modified AS 4100 formulae for unstrengthened aluminium RHS

Unlike cold-formed RHS shown in Figure 6.2(a), the aluminium RHS (shown in Figure 6.2(b)) does not have rounded corners. The bearing load can be transferred more directly to the webs of a sharp-corner aluminium section, which produces less load eccentricity to the web. Web buckling was found to be the dominant failure mode for aluminium RHS with a d/t_w ratio ranging from 17 to 68. The design formulae for web buckling are similar to those described in Section 6.2.5.1 by replacing r_{ext} with t_f.

$$R_{bb} = 2 \cdot b_b \cdot t_w \cdot f_y \cdot \alpha_c \qquad (6.11a)$$

$$b_b = b_s + 0.5 \cdot d + 1.5 \cdot t_f \qquad (6.11b)$$

α_c is the member slenderness reduction factor given in Figure 6.9 with a modified slenderness.

$$\lambda_n = k_e \cdot \sqrt{12} \cdot \left(\frac{d - 2 \cdot t_f}{t_w} \right) \cdot \sqrt{\frac{f_y}{250}} \qquad (6.11c)$$

Zhou et al. (2009) and Wu et al. (2011) showed that an effective length factor k_e of 1.05 gave good predictions of web buckling capacity.

6.3.5.2 Modified AS 4100 formulae for CFRP-strengthened aluminium RHS

The failure mode after type O and type I strengthening remains to be web buckling (see Figure 6.11(b) and (c)). Wu et al. (2011) proposed to use

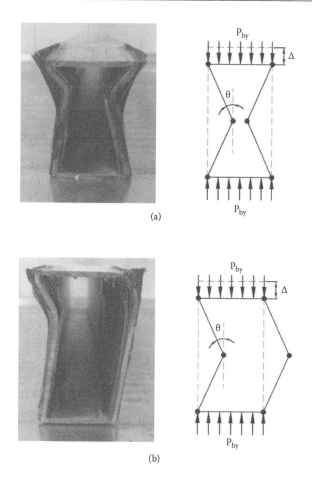

Figure 6.13 Plastic mechanism for aluminium RHS strengthened with type B. (a) Assumed inward mechanism. (b) Assumed sideway mechanism. (From Wu, C. et al., *Thin-Walled Structures*, 49(10), 1195–1207, 2011.)

Equation (6.11) with an effective buckling length factor k_e of 0.78 for type O strengthening and 0.65 for type I strengthening.

The failure mode after type B strengthening changes to web yielding, as shown in Figure 6.13. The plastic mechanism looks different compared with that for cold-formed RHS reported in Zhao and Hancock (1995). The assumed plastic mechanism for CFRP-strengthened RHS is shown in Figure 6.13, where P_{by} is the applied load and Δ is the deformation.

Both mechanisms in Figure 6.13 have the same expression of $\cos\theta$:

$$\cos\theta = 1 - \frac{\Delta}{(d/2)} \qquad (6.12)$$

The virtual change of the angle (θ):

$$\delta\theta = \frac{\delta\Delta}{(d/2)\sqrt{1-(1-\Delta/(d/2))^2}} \tag{6.13}$$

From the virtual work principle,

$$\delta W_{int} = \delta W_{ext} \tag{6.14}$$

i.e.,

$$4b_b\{M_p + M_p(1-\alpha_p^2)\}\delta\theta = 2P_{by}\delta\Delta \tag{6.15}$$

where b_b is the assumed mechanism length along the web given in Equation (6.11b), M_p is the plastic moment of a unit length $\left(f_y t_w^2/4\right)$, and α_p is defined (Zhao and Hancock 1993) in Equation (6.16).

$$\alpha_p = \frac{P_{by}/2}{b_b t_w f_y} \tag{6.16}$$

Equation (6.15) can be rewritten as

$$(2-\alpha_p^2)t_w = \alpha_p(d/2)\sqrt{1-(1-\Delta/(d/2))^2}$$

When $\Delta/(d/2) \ll 1$,

$$\sqrt{1-(1-\Delta/(d/2))^2} \approx 0$$

Therefore,

$$\alpha_p = \sqrt{2} \tag{6.17}$$

The web bearing yield capacity (R_{by}) becomes

$$R_{by} = 2\sqrt{2}(b_s + 0.5d + 1.5t_w)t_w f_y \tag{6.18}$$

6.3.5.3 AS/NZS 1664.1 formula for web bearing capacity of aluminium RHS

The Australian/New Zealand Standard (AS/NZS) 1664.1 (Standards Australia 1997) for aluminium structures has adopted the web crippling design rules from the AA specification (AA 2005). These design rules were based on the work done by Sharp and Jaworski (1991) and Sharp (1993). No distinction is made between web buckling capacity and web yielding capacity. The web bearing capacity (R_b) formulae for aluminium RHS are summarised below for the convenience of readers.

$$R_b = 2 \times \frac{1.2C_{wa}(b_s + C_{w2})}{C_{wb}} \tag{6.19a}$$

$$C_{wa} = t_w^2 \left(0.46f_y + 0.02\sqrt{E_{Al}f_y} \right) \tag{6.19b}$$

$$C_{wb} = 10 \text{ mm} \tag{6.19c}$$

$$C_{w2} = 33 \text{ mm} \tag{6.19d}$$

Equation (6.19) can be rewritten as

$$R_b = 0.24(b_s + 33)t_w^2 \left(0.46f_y + 0.02\sqrt{E_{Al}f_y} \right) \tag{6.20}$$

in which b_s is the bearing length (in mm), t_w is the web thickness, E_{Al} is the modulus of elasticity, and f_y is the yield stress of aluminium RHS.

6.3.5.4 Modified AS/NZS 1664.1 formula for web bearing capacity of CFRP-strengthened aluminium RHS

Since there is no distinction between web buckling and web yielding in AS/NZS 1664.1, the design formulae are considered together for type O, type I, and type B strengthening. Wu et al. (2011) proposed an equivalent thickness method to modify Equation (6.20) for CFRP-strengthened aluminium RHS. This method considers the constraining effect of CFRP attachment based on observation from the experiments that CFRP reinforcement engaged in load carrying with the web during the loading process (Wu et al. 2012). Therefore, the constrain effect of CFRP strengthening on the bearing capacity of the web can be reflected by increased wall thickness. The concept is illustrated in Figure 6.14.

It is assumed that the total applied load per unit web length (P_b) is shared by that carried by aluminium (P_{Al}) and that carried by CFRP (P_{CFRP}) with the same shortening of the web (Δ).

$$P_b = P_{Al} + P_{CFRP} \tag{6.21}$$

$$E_{Al}t_{we}\left(\frac{\Delta}{d}\right) = E_{Al}t_w\left(\frac{\Delta}{d}\right) + E_{CFRP}t_{CFRP, total}\left(\frac{\Delta}{d}\right) \tag{6.22}$$

Equation (6.22) can be simplified as

$$t_{we} = t_w + t_{CFRP, total}\frac{E_{CFRP}}{E_{Al}} \tag{6.23a}$$

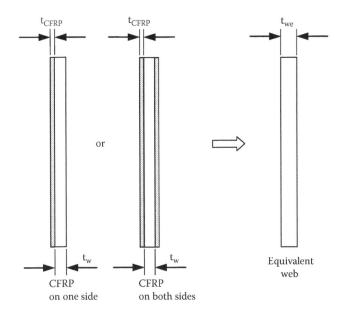

Figure 6.14 Schematic view of equivalent thickness method for aluminium RHS strengthened with either type O, type I, or type B.

$$t_{CFRP, total} = t_{CFRP} \text{ for type O and type I strengthening} \qquad (6.23b)$$

$$t_{CFRP, total} = 2t_{CFRP} \text{ for type B strengthening} \qquad (6.23c)$$

where t_{we} is the equivalent web thickness, t_w is the web thickness, t_{CFRP} is the thickness of one layer of CFRP plate, E_{Al} and E_{CFRP} are the moduli of elasticity of aluminium and CFRP plate, respectively.

After the equivalent thickness is substituted into Equation (6.20), the modified bearing capacity of CFRP-strengthened RHS can be achieved. However, type O and type I strengthening have the same equivalent web thickness according to their strengthening configurations, which means they will have the same nominal bearing capacity from Equation (6.20). Type B has twice the thickness of CFRP because CFRP is applied on both sides of the web. A modification factor (α) is introduced to take into account the difference in strengthening schemes. Equation (6.20) can be rewritten for CFRP-strengthened aluminium RHS as Equation (6.24a). The value of α was calibrated using the test data (Wu et al. 2011) and shown in Equation (6.24b) to Equation (6.24d).

$$R_b = 0.24(b_s + 33)t_{we}^2(0.46f_y + 0.02\sqrt{E_{Al}f_y})\alpha \qquad (6.24a)$$

$$\alpha = 0.75 \text{ for type O} \qquad\qquad (6.24\text{b})$$

$$\alpha = 1.0 \text{ for type I} \qquad\qquad (6.24\text{c})$$

$$\alpha = 0.75 \text{ for type B} \qquad\qquad (6.24\text{d})$$

The validity range of d/t_w is between 17 and 68.

6.4 LITESTEEL BEAMS

6.4.1 Types of strengthening

Zhao and Al-Mahaidi (2009) carried out an experimental program using CFRP to strengthen LiteSteel beams. The strengthening scheme is very much the same as that described in previous sections for RHS. Three types of strengthening methods (type O, type I, and type B) are shown in Figure 6.15.

6.4.2 Failure modes and behaviour

The progressive failure of each specimen was captured by taking photos at various load levels. An example is shown in Figure 6.16 for the type I method. Typical failure modes of each type of strengthening are shown in Figure 6.17. Web buckling was clearly observed for LSB without CFRP (see Figure 6.17(a)). With type O strengthening, CFRP debonding was evident, as shown in Figure 6.17(b). Once the debonding occurs the load starts to

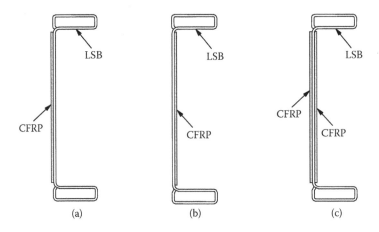

Figure 6.15 Types of strengthening for LiteSteel beams: (a) type O, (b) type I, and (c) type B. (From Zhao, X.L., and Al-Mahaidi, R., *Thin-Walled Structures*, 47(10), 1029–1036, 2009.)

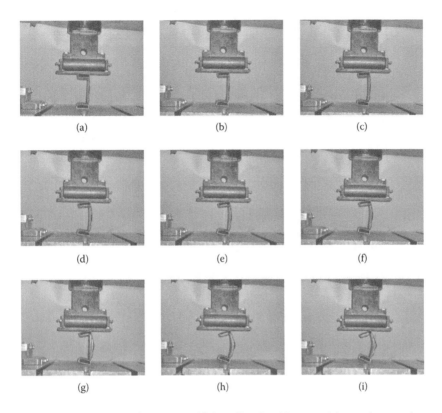

Figure 6.16 An example of progress of failure (LiteSteel beams with type I strengthening). (Courtesy of I. Tang, Monash University, Australia.)

drop, as shown in Figure 6.18 (type O curve). The CFRP debonding on the outer surface of the web was less evident for the type B method because of the involvement of the CFRP on the inner surface. In specimens strengthened using type I and type B methods, CFRP debonding also occurred on the inner surface. However, the CFRP on the inner surface continued to carry load even though partial debonding had occurred. This can be seen from the load-displacement curves shown in Figure 6.18.

6.4.3 Increased capacity

The increase in bearing capacity is shown in Figure 6.19(a) by plotting the ratio (CFRP-strengthened LSB to bare LSB) against the overall web depth-to-thickness ratio. It is evident that a significant increase in bearing capacity is obtained in the case of the type O method. The increase ranges from three to five times the strength of the bare LSB. The plot shows that significant increase in bearing capacity is obtained for larger d/t_w ratios.

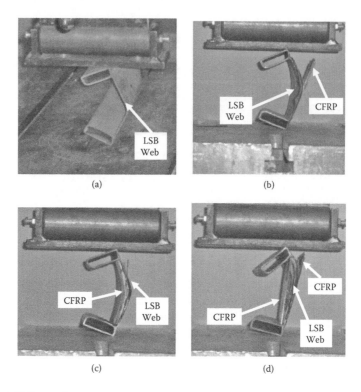

Figure 6.17 Typical failure modes for each strengthening type (LiteSteel beams): (a) without CFRP, (b) type O, (c) type I, and (d) type B. (Courtesy of I. Tang, Monash University, Australia.)

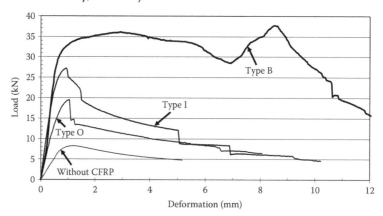

Figure 6.18 Typical load versus deformation curves: LiteSteel beams (d = 125 mm, b = 45 mm, d_f = 15 mm, and t_w = 1.6 mm) with various types of strengthening. (Adapted from Zhao, X.L., and Al-Mahaidi, R., *Thin-Walled Structures*, 47(10), 1029–1036, 2009.)

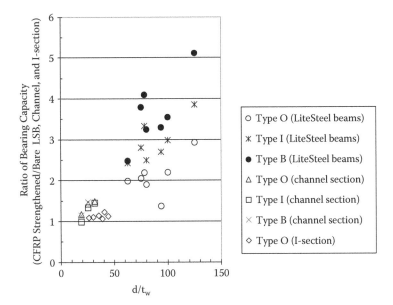

Figure 6.19 Increase in bearing capacity for LiteSteel beams, channel section, and I-section beams.

Figure 6.19 also shows that better strengthening can be achieved by applying CFRP on the inner side or on both sides of the section than by applying CFRP on the outer side alone.

6.4.4 Design formulae

6.4.4.1 Modified AS 4100 formulae for unstrengthened LiteSteel beams

LiteSteel beams have a unique shape (shown in Figure 6.2(c)) when compared with RHS or I-section. Therefore, the load dispersion to the web is different from that given in Figure 6.8 for RHS or I-section. It can be assumed that the bearing load is distributed to the web directly, as shown in Figure 6.20. The design formulae for web buckling are similar (but having only one web) to those described in Section 6.2.5.1 by using a new b_b dimension (see Figure 6.20) and web depth $(d - 2d_f)$.

$$R_{bb} = b_b \cdot t_w \cdot f_y \cdot \alpha_c \qquad (6.25a)$$

$$b_b = b_s + 0.5 \cdot (d - 2d_f) \qquad (6.25b)$$

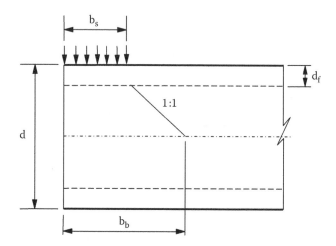

Figure 6.20 Load dispersion for LiteSteel beams subject to end bearing force.

α_c is the member slenderness reduction factor given in Figure 6.9 with a modified slenderness.

$$\lambda_n = k_e \cdot \sqrt{12} \cdot \left(\frac{d - 2 \cdot d_f}{t_w} \right) \cdot \sqrt{\frac{f_y}{250}} \qquad (6.25c)$$

Zhao and Al-Mahaidi (2009) showed that an effective length factor k_e of 1.0 gave reasonable predictions of web buckling capacity.

6.4.4.2 Modified AS 4100 formulae for CFRP-strengthened LiteSteel beams

Zhao and Al-Mahaidi (2009) presented design formulae for CFRP-strengthened LiteSteel beams using type B (see Figure 6.15(c)), which are the same as Equation (6.25) with an effective length factor k_e of 0.5. It can be proven that an effective length factor of 0.65 is suitable for type O and 0.55 is suitable for type I. Hence,

$k_e = 0.65$ for type O strengthening $\qquad (6.26a)$

$k_e = 0.55$ for type I strengthening $\qquad (6.26b)$

$k_e = 0.50$ for type B strengthening $\qquad (6.26c)$

6.5 OPEN SECTIONS

6.5.1 Types of strengthening

Zhao (2009) studied the influence of CFRP strengthening on the web bearing capacity of channel and I-section beams. For channel sections, three types of strengthening methods (O, I, and B) are adopted, as described previously in this chapter for RHS and LSB. They are shown in Figure 6.21(a) to (c). For I-section, only type B is investigated because it is a symmetric section, as shown in Figure 6.21(d).

6.5.2 Failure modes and increased capacity

The failure modes are shown in Figure 6.22(a) for channel sections and in Figure 6.22(b) for I-sections. Web buckling is clearly shown in Figure 6.22(a)(i) for a bare channel section. Figure 6.22(a)(ii) demonstrates that less buckling of the section occurs due to CFRP strengthening. CFRP delamination happens on the outer web of the channel section. A similar phenomenon is observed for I-sections, as shown in Figure 6.22(b).

Increased capacity due to CFRP strengthening is presented in Figure 6.19 for channel and I-sections. It can be seen that more increase in web buckling capacity is achieved for slender (i.e., larger d/t_w ratio) steel sections. The increase is up to 50% for channel sections and up to 30% for I-sections. For channel sections the type O method is not very effective because of CFRP delamination. Type I and type B achieved similar results. It seems that applying CFRP to the inner web of a channel section alone is sufficient for web buckling strengthening.

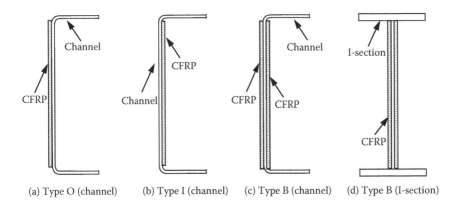

(a) Type O (channel)   (b) Type I (channel)   (c) Type B (channel)   (d) Type B (I-section)

Figure 6.21 Type of strengthening for channel and I-sections.

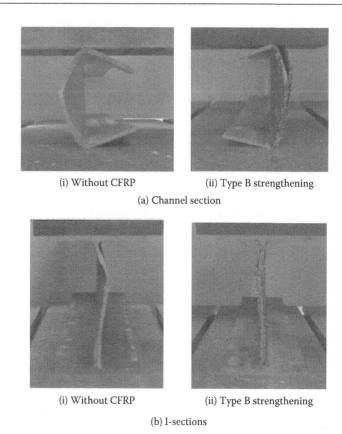

(i) Without CFRP   (ii) Type B strengthening

(a) Channel section

(i) Without CFRP   (ii) Type B strengthening

(b) I-sections

Figure 6.22 Typical failure modes for each strengthening type (channel and I-sections). (Courtesy of M. O'Dwyer, Monash University, Australia.)

6.5.3 Design formulae

6.5.3.1 Modified Young and Hancock (2001) formulae for CFRP-strengthened channel section

Young and Hancock (2001) proposed equations for web buckling capacity of channel sections subject to end bearing forces. They can be rewritten as follows using the symbols defined in this chapter (see Figure 6.2(b)):

$$R_{bb,\,channel} = \frac{f_y t_w^2}{4} \cdot \left(\frac{b_s + 0.3d}{r_i + 0.5t_w} \right) \cdot \left[1.44 - 0.0133 \left(\frac{h}{t} \right) \right] \qquad (6.27a)$$

$$h = d - 2 \cdot t_w - 2 \cdot r_i \qquad (6.27b)$$

in which f_y is the yield stress, t_w is the web thickness, r_i is the inner corner radius, b_s is the bearing length, and d is the overall depth of the channel section.

The above equations were derived through a combination of theoretical and empirical analyses (Young and Hancock 2001). The use of the CFRP increases the web buckling capacity of channel sections as shown in Figure 6.19. There is not much difference in the increased capacity for three types of strengthening methods (type O, type I, and type B). The increase seems to be linear as d/t_w increases. A simple approach to consider the influence of CFRP is to introduce an amplification factor (K_{CFRP}), which can be derived from Figure 6.19 using a regression analysis. The following relationship was found by Zhao (2009) for the amplification factor.

$$K_{CFRP} = 0.5 + 0.03 \cdot \frac{d}{t_w} \tag{6.28}$$

Hence the web buckling capacity of CFRP-strengthened channel sections subject to end bearing force becomes

$$R_{bb, CFRP + channel} = K_{CFRP} \cdot R_{bb, channel} \tag{6.29}$$

where $R_{bb,channel}$ is determined from Equation (6.27), whereas K_{CFRP} is given in Equation (6.28). The d/t_w range investigated was from 20 to 40. When d/t_w is greater than 40, the value of d/t_w of 40 may be adopted to be conservative.

6.5.3.2 Modified AS 4100 formulae for CFRP-strengthened I-section

AS 4100 (Standards Australia 1998) has equations for unstrengthened I-sections subject to end bearing forces. The dispersion of bearing load is shown in Figure 6.8. The design formulae for web buckling are similar (but having only one web) to those described in Section 6.2.5.1 by replacing r_{ext} by t_f.

$$R_{bb} = b_b \cdot t_w \cdot f_y \cdot \alpha_c \tag{6.30a}$$

$$b_b = b_s + 0.5 \cdot d + 1.5 \cdot t_f \tag{6.30b}$$

α_c is the member slenderness reduction factor given in Figure 6.9 with a modified slenderness:

$$\lambda_n = k_e \cdot \sqrt{12} \cdot \left(\frac{d - 2 \cdot t_f}{t_w} \right) \cdot \sqrt{\frac{f_y}{250}} \tag{6.30c}$$

For an unstrengthened I-section subject to end bearing force, the effective buckling length factor k_e is given as 0.7 in AS 4100. For a CFRP-strengthened I-section using type B, a reduced k_e of 0.5 was found suitable (Zhao 2009) for the design of such an I-section subject to end bearing force.

6.6 DESIGN EXAMPLES

All the design formulae are summarised in Table 6.1 for the convenience of readers. The design capacity factor (ϕ) will be described within the design examples.

6.6.1 Example I (cold-formed RHS)

A cold-formed RHS 100 × 50 × 2.5 is subjected to an end bearing force (R^*) of 40 kN (see Figure 6.8). The force is applied over the full width of the RHS and for a length of 25 mm along the RHS. Assume that the yield stress is 355 N/mm². Check the bearing capacity of the beam. If the RHS is not sufficient, choose one CFRP strengthening type to increase the bearing capacity. The design capacity factor (ϕ) can be taken as 0.9 as in AS 4100.

6.6.1.1 Solution according to AS 4100 given in Section 6.2.5 for unstrengthened RHS

1. Dimensions and properties:
   $d = 100$ mm
   $b = 50$ mm
   $t_w = 2.5$ mm
   $b_s = 25$ mm
   From Equation (6.5b), the external corner radius of the RHS becomes
   $r_{ext} = 2\, t_w = 2 \times 2.5 = 5$ mm
   From Equation (6.4) the dimension b_b can be determined as
   $b_b = b_s + 0.5d + 1.5 \cdot r_{ext} = 25 + 0.5 \times 100 + 1.5 \times 5 = 82.5$ mm
   $f_y = 355$ N/mm²
2. Web bearing buckling versus web bearing yield:
   Since
   $$\left(\frac{d - 2 \cdot r_{ext}}{t_w} \right) \cdot \sqrt{\frac{f_y}{250}} = \left(\frac{100 - 2 \times 5}{2.5} \right) \sqrt{\frac{355}{250}} = 43 > 36$$

   web bearing buckling governs (see Equation 6.8a), i.e.,
   $R_b = R_{bb}$
3. Web buckling capacity:
   From Equation (6.2),
   $R_{bb} = 2 \cdot b_b \cdot t_w \cdot f_y \cdot \alpha_c$

Table 6.1 Summary of design approaches

Cross section type	Design basis	Without CFRP	Type O	Type I	Type B
Cold-formed RHS (web buckling governs)[a]	Modified AS 4100	Web buckling formula (see Equations (6.2), (6.4), and (6.6)) with k_e of 1.1	Web bearing formula (see Equations (6.1) to (6.7)) with k_e of 0.8	Web yielding formula with α_p of 0.32 (see Equation (6.9))	
Cold-formed RHS (web yielding governs)[a]	Modified AS 4100	Web yielding formula with α_p given in Equation (6.7)	Web yielding formula with α_p of 0.32 and f_u to replace f_y (see Equation (6.10))		
Aluminium RHS	Modified AS 4100	Web buckling formula (see Equation (6.11)) with k_e of 1.05	Web buckling formula (see Equation (6.11)) with k_e of 0.78	Web buckling formula (see Equation (6.11)) with k_e of 0.65	Web yielding formula with α_p of $\sqrt{2}$ and b_b of $(b_s + 0.5d + 1.5t_w)$ (see Equation (6.18))
Aluminium RHS	Modified AS/NZS 1664	Empirical equation given in Equation (6.20)	Equivalent thickness method (see Equation (6.24)) with t_{we} given in Equation (6.23)		
LiteSteel beams	Modified AS 4100	Web buckling formula (see Equation (6.25)) with k_e of 1.0	Web buckling formula (see Equation (6.25)) with k_e of 0.65	Web buckling formula (see Equation (6.25)) with k_e of 0.55	Web buckling formula (see Equation (6.25)) with k_e of 0.5
Channel section	Modified Young and Hancock (2001)	Empirical equation given in Equation (6.27)	Empirical equation (Equation (6.29)) with an amplification factor given in Equation (6.28)		
I-Section	Modified AS 4100	Web buckling formula (Equation (6.30)) with k_e of 0.7	N/A	N/A	Web buckling formula (Equation (6.30)) with k_e of 0.5

[a] To determine if a cold-formed RHS without CFRP fails in web buckling or web yielding, see Equation (6.8).

α_c can be determined from Figure 6.9 using the λ_n defined in Equation (6.6) with k_e of 1.1:

$$\lambda_n = 1.1 \times \sqrt{12} \cdot \left(\frac{d - 2 \cdot r_{ext}}{t_w} \right) \cdot \sqrt{\frac{f_y}{250}} = 1.1 \times 3.5 \times 43 = 165$$

Hence, $\alpha_c \approx 0.27$.
$R_b = R_{bb} = 2 \cdot b_b \cdot t_w \cdot f_y \cdot \alpha_c = 2 \times 82.5 \times 2.5 \times 355 \times 0.27 = 39{,}538$ kN = 39.5 kN
$\phi R_b = 0.9 \times 39.5 = 35.6$ kN < $R^* = 40$ kN
The cold-formed RHS $100 \times 50 \times 2.5$ is not satisfactory.

6.6.1.2 Solution according to modified AS 4100 given in Section 6.2.5 for CFRP-strengthened RHS

Choose type O strengthening method.
    According to Table 6.1, the web bearing formula (Equations (6.1) to (6.7)) with k_e of 0.8 should be used.
    α_c can be determined from Figure 6.9 using the λ_n defined in Equation (6.6) with k_e of 0.8:

$$\lambda_n = 0.8 \times \sqrt{12} \cdot \left(\frac{d - 2 \cdot r_{ext}}{t_w} \right) \cdot \sqrt{\frac{f_y}{250}} = 0.8 \times 3.5 \times \left(\frac{100 - 2 \times 5}{2.5} \right) \cdot \sqrt{\frac{355}{250}} = 120$$

Hence, $\alpha_c \approx 0.46$.
    Web buckling capacity:

$$R_{bb} = 2 \cdot b_b \cdot t_w \cdot f_y \cdot \alpha_c = 2 \times 82.5 \times 2.5 \times 355 \times 0.46 = 67{,}361 \text{ N}$$
$$= 67.4 \text{ kN}$$

From Equation (6.7),

$$k_s = \frac{2 \cdot r_{ext}}{t_w} - 1 = \frac{2 \times 5}{2.5} - 1 = 3$$

$$\alpha_p = \sqrt{(2 + k_s^2)} - k_s = \sqrt{(2 + 3^2)} - 3 = 0.317$$

From Equation (6.3), web yielding capacity:

$$R_{by} = 2 \cdot b_b \cdot t_w \cdot f_y \cdot \alpha_p = 2 \times 82.5 \times 2.5 \times 355 \times 0.317 = 46{,}421 \text{ N}$$
$$= 46.4 \text{ kN}$$

From Equation (6.1), web bearing capacity:

$$R_b = \min \{R_{bb}, R_{by}\} = \min \{67.4, 46.4\} = 46.4 \text{ kN}$$

$$\phi R_b = 0.9 \times 46.4 = 41.8 \text{ kN} > R^* = 40 \text{ kN}$$

The strengthened RHS using type O is satisfactory.

6.6.2 Example 2 (aluminium RHS)

An aluminium RHS $100 \times 50 \times 2$ is subjected to an end bearing force (R^*) of 20 kN (see Figure 6.8). The force is applied over the full width of the RHS and for a length of 25 mm along the RHS. Assume that the yield stress is 235 N/mm² and modulus of elasticity is 70,000 N/mm². The modulus of elasticity for the CFRP plate is 165,000 N/mm² and the thickness of the CFRP plate is 1.2 mm. Check the bearing capacity of the beam. If the RHS is not sufficient, choose one CFRP strengthening type to increase the bearing capacity. The design capacity factor (ϕ) can be taken as 0.9, as in AS 4100 and AS 1664.1.

6.6.2.1 Solution according to modified AS 4100 given in Section 6.3.5

a. Aluminium RHS without CFRP:
  1. Dimensions and properties:
     $d = 100$ mm
     $b = 50$ mm
     $t_w = 2$ mm
     $t_f = 2$ mm
     $b_s = 25$ mm
     $t_{CFRP} = 1.2$ mm
     From Equation (6.11b) the dimension b_b can be determined as
     $b_b = b_s + 0.5 \cdot d + 1.5 \cdot t_f = 25 + 0.5 \times 100 + 1.5 \times 2 = 78$ mm
     $f_y = 235$ N/mm²
     $E_{Al} = 70{,}000$ N/mm²
     $E_{CFRP} = 165{,}000$ N/mm²
  2. Web bearing buckling capacity:
     From Equation (6.11a)
     $R_{bb} = 2 \cdot b_b \cdot t_w \cdot f_y \cdot \alpha_c$
     α_c can be determined from Figure 6.9 using the λ_n defined in Equation (6.11c) with k_e of 1.05:

     $$\lambda_n = 1.05 \times \sqrt{12} \cdot \left(\frac{d - 2 \cdot t_f}{t_w} \right) \cdot \sqrt{\frac{f_y}{250}}$$

     $$= 1.05 \times 3.5 \times \left(\frac{100 - 2 \times 2}{2} \right) \cdot \sqrt{\frac{235}{250}} = 171$$

     Hence, $\alpha_c \approx 0.25$.
     $R_{bb} = 2 \cdot b_b \cdot t_w \cdot f_y \cdot \alpha_c = 2 \times 78 \times 2 \times 235 \times 0.25 = 18{,}330$ N
     $\quad = 18.3$ kN
     $\phi R_{bb} = 0.9 \times 18.3 = 16.5$ kN $< R^* = 20$ kN
     The aluminium RHS $100 \times 50 \times 2$ is not satisfactory.
b. Aluminium RHS with CFRP strengthening:
   Choose type O strengthening method.

According to Section 6.3.5.2, the web buckling capacity for the type O strengthening method can be determined using Equation (6.11) with an effective buckling length factor k_e of 0.78.

α_c can be determined from Figure 6.9 using the λ_n defined in Equation (6.11c) with k_e of 0.78:

$$\lambda_n = 0.78 \times \sqrt{12} \cdot \left(\frac{d - 2 \cdot t_f}{t_w} \right) \cdot \sqrt{\frac{f_y}{250}}$$

$$= 0.78 \times 3.5 \times \left(\frac{100 - 2 \times 2}{2} \right) \cdot \sqrt{\frac{235}{250}} = 127$$

Hence, $\alpha_c \approx 0.42$.

$R_{bb} = 2 \cdot b_b \cdot t_w \cdot f_y \cdot \alpha_c = 2 \times 78 \times 2 \times 235 \times 0.42 = 30{,}794 \text{ N} = 30.8 \text{ kN}$

$\phi R_{bb} = 0.9 \times 30.8 = 27.7 \text{ kN} > R^* = 20 \text{ kN}$

The type O strengthened aluminium RHS 100 × 50 × 2 is satisfactory.

6.6.2.2 Solution according to modified AS 1664.1 given in Section 6.3.5

a. Aluminium RHS without CFRP:

From Equation (6.20), web bearing capacity of aluminium RHS:

$$R_b = 0.24(b_s + 33)t_w^2 \left(0.46 f_y + 0.02 \sqrt{E f_y} \right)$$

$$= 0.24 \times (25 + 33) \times 2^2 \times \left(0.46 \times 235 + 0.02 \times \sqrt{70{,}000 \times 235} \right)$$

$$= 10{,}536 \text{ N} = 10.5 \text{ kN}$$

$\phi R_b = 0.9 \times 10.5 = 9.45 \text{ kN} < R^* = 20 \text{ kN}$

The aluminium RHS 100 × 50 × 2 is not satisfactory.

b. Aluminium RHS with CFRP strengthening:

Choose type O strengthening method.

According to Equation (6.23), the equivalent thickness becomes

$$t_{we} = t_w + t_{CFRP,total} \frac{E_{CFRP}}{E_{Al}} = 2 + 1.2 \times \frac{165{,}000}{70{,}000} = 4.83 \text{ mm}$$

From Equations (6.24a) and (6.24b):

$$R_b = 0.24(b_s + 33)t_{we}^2 \left(0.46 f_y + 0.02 \sqrt{E f_y} \right) \alpha$$

$$= 0.24 \times (25 + 33) \times 4.83^2 \times \left(0.46 \times 235 + 0.02 \times \sqrt{70{,}000 \times 235} \right) \times 0.75$$

$$= 46{,}084 \text{ N} = 46.1 \text{ kN}$$

$\phi R_b = 0.9 \times 46.1 = 41.2 \text{ kN} > R^* = 20 \text{ kN}$
The type O strengthened aluminium RHS $100 \times 50 \times 2$ is satisfactory.

6.6.3 Example 3 (LiteSteel beams)

An LSB $125 \times 45 \times 15.2 \times 2$ ($d \times b \times d_f \times t_w$) is subjected to an end bearing force (R^*) of 15 kN (see Figure 6.20). The force is applied over the full width of the LSB and for a length of 25 mm along the LSB. Assume that the yield stress is 380 N/mm². Check the bearing capacity of the beam. If the LSB is not sufficient, choose one CFRP strengthening type to increase the bearing capacity. The design capacity factor (ϕ) can be taken as 0.9 as in AS 4100.

6.6.3.1 Solution according to modified AS 4100 given in Section 6.4.4 for unstrengthened LSB

1. Dimensions and properties:
   $d = 125$ mm
   $b = 45$ mm
   $d_f = 15.2$ mm
   $t_w = 2$ mm
   $b_s = 25$ mm
   $f_y = 380$ N/mm²
   From Equation (6.25b) the dimension b_b can be determined as
   $b_b = b_s + 0.5 \cdot (d - 2d_f) = 25 + 0.5 \times (125 - 2 \times 15.2) = 72.3$ mm
2. Web bearing buckling capacity:
   From Equation (6.25a),
   $R_{bb} = b_b \cdot t_w \cdot f_y \cdot \alpha_c$
   α_c can be determined from Figure 6.9 using the λ_n defined in Equation (6.25c) with k_e of 1.0:

   $$\lambda_n = k_e \cdot \sqrt{12} \cdot \left(\frac{d - 2 \cdot d_f}{t_w} \right) \cdot \sqrt{\frac{f_y}{250}}$$

   $$= 1.0 \times \sqrt{12} \cdot \left(\frac{125 - 2 \times 15.2}{2} \right) \cdot \sqrt{\frac{380}{250}} = 202$$

   Hence, $\alpha_c \approx 0.18$.
   Web buckling capacity becomes
   $R_{bb} = b_b \cdot t_w \cdot f_y \cdot \alpha_c = 72.3 \times 2 \times 380 \times 0.18 = 9891 \text{ N} = 9.89 \text{ kN}$
   $\phi R_b = 0.9 \times 9.89 = 8.90 \text{ kN} < R^* = 15 \text{ kN}$
   The unstrengthened LSB $125 \times 45 \times 15.2 \times 2$ is not satisfactory.

6.6.3.2 Solution according to modified AS 4100 given in Section 6.4.4 for CFRP-strengthened LSB

Choose type O strengthening.

According to Section 6.4.4.2, the web buckling capacity for the type O strengthening method can be determined using Equation (6.25) with an effective buckling length factor k_e of 0.65.

α_c can be determined from Figure 6.9 using the λ_n defined in Equation (6.25c) with k_e of 0.65:

$$\lambda_n = k_e \cdot \sqrt{12} \cdot \left(\frac{d - 2 \cdot d_f}{t_w} \right) \cdot \sqrt{\frac{f_y}{250}} = 0.65 \times \sqrt{12} \cdot \left(\frac{125 - 2 \times 15.2}{2} \right) \cdot \sqrt{\frac{380}{250}} = 131$$

Hence, $\alpha_c \approx 0.40$.

Web buckling capacity becomes

$$R_{bb} = b_b \cdot t_w \cdot f_y \cdot \alpha_c = 72.3 \times 2 \times 380 \times 0.40 = 21,979 \text{ N} = 22.0 \text{ kN}$$

$$\phi R_b = 0.9 \times 22.0 = 19.8 \text{ kN} > R^* = 15 \text{ kN}$$

The type O strengthened LSB $125 \times 45 \times 15.2 \times 2$ is satisfactory.

6.7 FUTURE WORK

Most of the work completed so far was on steel sections subject to end bearing forces. There is a need to investigate FRP strengthening of web crippling in other materials (e.g., aluminium and stainless steel) subject to various load conditions, such as end-two-flange (ETF), interior-two-flange (ITF), end-one-flange (EOF) and interior-one-flange (IOF) loadings (Islam and Young 2011, 2012a, 2012b, 2013). There is a lack of detailed FE simulation of web crippling with FRP strengthening (Fernando et al. 2009). The proposed design formulae are rather simplistic. More sophisticated theoretical models are needed to predict more accurately the web buckling capacity.

REFERENCES

AA. 2005. *Aluminum design manual.* Washington, DC: Aluminum Association.

ASI. 1999. *Design capacity tables for structural steel: Hollow sections.* Vol. 2. Sydney: Australian Steel Institute.

Fernando, D., Yu, T., Teng, J.G., and Zhao, X.L. 2009. CFRP strengthening of rectangular steel tubes subjected to end bearing loads: Effect of adhesive properties and finite element modelling. *Thin-Walled Structures*, 47(10), 1020–1028.

Hancock, G.J. 1994. *Design of cold-formed steel structures*. Sydney: Australian Institute of Steel Construction.

Islam, S.M.Z., and Young, B. 2011. FRP strengthened aluminium tubular sections subjected to web crippling. *Thin-Walled Structures*, 49(11), 1392–1403.

Islam, S.M.Z., and Young, B. 2012a. Web crippling of aluminium tubular structural members strengthened by CFRP. *Thin-Walled Structures*, 59(10), 58–69.

Islam, S.M.Z., and Young, B. 2012b. Ferritic stainless steel tubular members strengthened with high modulus CFRP plate subjected to web crippling. *Journal of Constructional Steel Research*, 77(10), 107–118.

Islam, S.M.Z., and Young, B. 2013. Strengthening of ferritic stainless steel tubular structural members using FRP subjected to two-flange-loading. *Thin-Walled Structures*, 62(1), 179–190.

Packer, J.A. 1984. Web crippling of rectangular hollow sections. *Journal of Structural Engineering*, 110(10), 2357–2373.

Packer, J.A. 1987. Review of American RHS web crippling provision. *Journal of Structural Engineering*, 113(12), 2508–2513.

Packer, J.A., and Fear, C.E. 1991. Concrete-filled rectangular hollow section X and T connections. In *Proceedings of the 4th International Symposium on Tubular Structures*, Delft, Delft University Press, pp. 382–391.

Sharp, M.L. 1993. *Behavior and design of aluminum structures*. New York: McGraw-Hill.

Sharp, M.L., and Jaworski, A.P. 1991. Methodology for developing an engineering solution for web crippling. In *Proceedings of Mechanics Computing in 1990's and Beyond*, Columbus, OH, May 20–22, pp. 877–881.

Standards Australia. 1997. *Aluminium structures. Part 1. Limit state design*. AS/NZS 1664.1. Sydney: Standards Australia.

Standards Australia. 1998. *Steel structures*. AS 4100. Sydney: Standards Australia.

Wilkinson, T., Liu, P., Magpayo, J., and Nguyen, H. 2006b. Increasing the strength and stiffness of cold-formed hollow flange channel sections for web crippling. In *Proceedings of the 18th Specialty Conference on Cold-Formed Steel Structures*, Orlando, FL, October, pp. 119–132.

Wilkinson, T., Zhu, Y., and Yang, D. 2006a. Behaviour of hollow flange channel sections under concentrated loads. In *Proceedings of the 11th International Symposium on Tubular Structures*, Québec, Canada, August 31–September 2, pp. 187–194.

Wu, C., Zhao, X.L., and Duan, W.H. 2011. Design rules for web crippling of CFRP strengthened aluminium rectangular hollow sections. *Thin-Walled Structures*, 49(10), 1195–1207.

Wu, C., Zhao, X.L., Duan, W.H., and Phipatet, P. 2012. Improved end bearing capacities of sharp corner aluminum tubular sections with CFRP strengthening. *International Journal of Structural Stability and Dynamics*, 12(1), 109–130.

Young, B., and Hancock, G.J. 2001. Design of cold-formed channels subjected to web crippling. *Journal of Structural Engineering*, 127(10), 1137–1144.

Young, B., and Zhou, F. 2008. Aluminum tubular sections subjected to web crippling. Part II. Proposed design equations. *Thin-Walled Structures*, 46(4), 352–361.

Zhao, X.L. 1999. Partially stiffened RHS sections under transverse bearing force. *Thin-Walled Structures*, 35(3), 193–204.

Zhao, X.L. 2009. Tests on CFRP strengthened open sections subjected to end bearing forces. Presented at Proceedings of the Second Asia-Pacific Conference on FRP in Structures (APFIS 2009), Seoul, December 9–11.

Zhao, X.L., and Al-Mahaidi, R. 2009. Web buckling of LiteSteel beams strengthened with CFRP subjected to end bearing forces. *Thin-Walled Structures*, 47(10), 1029–1036.

Zhao, X.L., Fernando, D., and Al-Mahaidi, R. 2006. CFRP strengthened RHS subjected to transverse end bearing force. *Engineering Structures*, 28(11), 1555–1565.

Zhao, X.L., and Hancock, G.J. 1992. Square and rectangular hollow sections subject to combined actions. *Journal Structural Engineering*, 118(3), 648–668.

Zhao, X.L., and Hancock, G.J. 1993. Experimental verification of the theory of plastic moment capacity of an inclined yield line under axial force. *Thin-Walled Structures*, 15(3), 209–233.

Zhao, X.L., and Hancock, G.J. 1995. Square and rectangular hollow sections under transverse end bearing force. *Journal of Structural Engineering*, 121(9), 1323–1329.

Zhao, X.L., and Phipatet, P. 2010. Tests on CFRP strengthened aluminium RHS subject to end bearing force. Presented at Proceedings of the 5th International Conference on Composites in Civil Engineering, Beijing, September 27–29.

Zhao, X.L., Wilkinson, T., and Hancock, G.J. 2005. *Cold-formed tubular members and connections*. Oxford: Elsevier.

Zhou, F., and Young, B. 2008. Aluminum tubular sections subjected to web crippling. Part I. Tests and finite element analysis. *Thin-Walled Structures*, 46(4), 339–351.

Zhou, F., and Young, B. 2010. Web crippling of aluminium tubes with perforated webs. *Engineering Structures*, 32(5), 1397–1410.

Zhou, F., Young, B., and Zhao, X.L. 2009. Tests and design of aluminum tubular sections subjected to concentrated bearing load. *Journal of Structural Engineering*, 135(7), 806–817.

Chapter 7

Enhancement of fatigue performance

7.1 GENERAL

Large amounts of steel structures in road and railway infrastructure, mining, transportation, and recreation industries are subjected to fatigue loading. There is potential to apply fibre-reinforced polymer (FRP) to strengthen ageing metallic structures to extend their fatigue life. Some field applications are given in Chapter 1.

This chapter first describes various methods of strengthening adopted by researchers. Five strengthening schemes are defined for steel plate application. Examples of improvement in fatigue performance due to FRP strengthening are then given. Discussions are made on key parameters influencing the fatigue strengthening. It continues to introduce several methods to capture fatigue crack propagation in an FRP-steel system. More details are given about the beach marking method. Two approaches are presented to predict the fatigue life for centre-cracked tensile (CCT) steel plates strengthened by multiple layers of carbon fibre-reinforced polymer (CFRP) sheet. One is the boundary element method (BEM), and the other is the fracture mechanics approach. Both methods gave reasonable prediction of fatigue life for such a CFRP–steel composite system. Finally, derivation of the stress intensity factor (SIF) for CCT steel plates strengthened by CFRP is carried out. The strengthening of welded connections is not covered in this chapter since very limited results are available at the present time.

7.2 METHODS OF STRENGTHENING

Studies on fatigue crack propagation in steel members and connections strengthened with FRP were carried out by many researchers. Some examples are shown in Figure 7.1. They can be grouped into four major categories: (1) steel plates (e.g., Colombi et al. 2003, Jones and Civjan 2003, Suzuki 2004), (2) I-section beams (e.g., Tavakkolizadeh and Saadatmanesh

2003, Jiao et al. 2012), (3) welded attachments (e.g., Nakamura et al. 2009, Chen et al. 2012), and (4) welded tubular joints (e.g., Nadauld and Pantelides 2007, Xiao and Zhao 2012).

Steel plate specimens represent an idealised boundary condition subjected to tension force. They are more suitable for fundamental study of crack propagation and an analytical approach to predict fatigue life. The other types are more directly applicable to a particular type of structure, e.g., I-section beams for steel girder bridges, welded attachments for plate

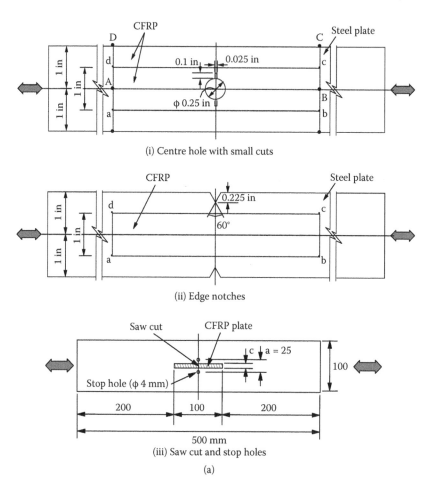

(i) Centre hole with small cuts

(ii) Edge notches

(iii) Saw cut and stop holes

(a)

Figure 7.1 Examples of fatigue strengthening. (a) Steel plates. (Adapted from Colombi, P. et al., *Fatigue and Fracture of Engineering Materials and Structures*, 26(1), 59–66, 2003; Jones, S.C., and Civjan, S.A., *Journal of Composites for Construction*, 7(4), 331–338, 2003; Suzuki, H., Experimental Study on Repair of Cracked Steel Member by CFRP Strip and Stop Hole, presented at Proceedings of the 11th European Conference on Composite Materials, Rhodes, Greece, May 31–June 3, 2004.)

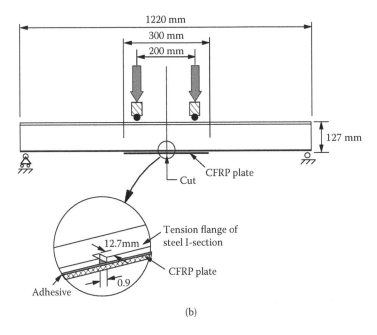

1220 mm

300 mm

200 mm

127 mm

CFRP plate

Cut

Tension flange of
steel I-section

12.7 mm

CFRP plate

0.9

Adhesive

(b)

Figure 7.1 (Continued) Examples of fatigue strengthening. (b) I-section beams. (Adapted from Tavakkolizadeh, M., and Saadatmanesh, H., *Journal of Structural Engineering*, 129(2), 186–196, 2003; Jiao, H. et al., Improving Fatigue Performance of CFRP Strengthened Steel Beams by Applying Vacuum Pressure in the Wet Layup of CFRP Woven Sheets, presented at Proceedings of the Third Asia-Pacific Conference on FRP in Structures (APFIS2012), Sapporo, Japan, February 2–4, 2012.)

girder and box girder bridges, welded cross-beams for undercarriages of motor vehicles and trailers, and welded tubular K-joints for truss joints of highway overhead sign structures.

Initial cracks are often used in the studies to either represent different damage levels or produce more consistent results for comparison. The commonly used initial cracks include a centre hole with small cuts, edge notches or a saw cut and stop holes on steel plates, an initial cut on the tension flange of I-beams, and different crack length in welded tubular joints.

Different researchers adopted different CFRP strengthening schemes for steel plate specimens by varying the bond width and location of the CFRP. Wu et al. (2012a) grouped commonly used strengthening schemes into five types, as shown in Figure 7.2. The terminology defined in Figure 7.2 (scheme (a) to scheme (e)) is used in this book when discussing strengthening of steel plates.

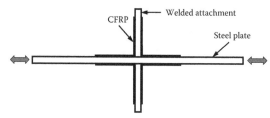

(i) Welded non-load carrying cruciform joints

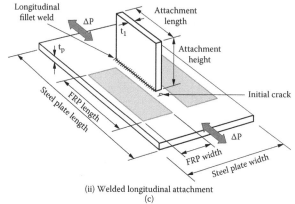

(ii) Welded longitudinal attachment
(c)

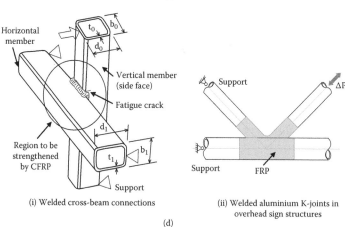

(i) Welded cross-beam connections

(ii) Welded aluminium K-joints in overhead sign structures

(d)

Figure 7.1 (Continued) Examples of fatigue strengthening. (c) Welded attachments. (Adapted from Nakamura, H. et al., *Thin-Walled Structures*, 47(10), 1059–1068, 2009; Chen, T., *International Journal of Structural Stability and Dynamics*, 12(1), 179–194, 2012.) (d) Welded tubular joints. (Adapted from Nadauld, J., and Pantelides, C.P., *Journal of Composites for Construction*, 11(3), 328–335, 2007; Xiao, Z.G., and Zhao, X.L., *International Journal of Structural Stability and Dynamics*, 12(1), 195–211, 2012.)

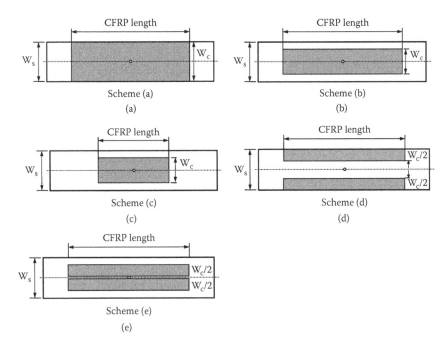

Figure 7.2 Schematic view of CFRP strengthening schemes used in the case of steel plates. (Adapted from Wu, C. et al., *Advances in Structural Engineering—An International Journal*, 15(10), 1801–1815, 2012.)

7.3 IMPROVEMENT IN FATIGUE PERFORMANCE

The fatigue phenomenon is characterised by a progressive degradation of strength under time-variant stresses (Grundy 2004). The fatigue life (or fatigue resistance) of a structure includes a crack initiation phase and a crack propagation phase. The fatigue life of welded connections depends on the connection detail and applied cyclic loading spectrum. It also depends on other factors, such as stress range applied, stress ratio, residual stress, and crack propagation rate. In an FRP-steel composite system the stress range in steel will decrease because some load is shared by FRP. The stress ratio may decrease because of the compressive stress field created by FRP, especially by prestressed FRP. The application of FRP to fatigue strengthening in civil engineering structures affects mainly the crack propagation phase rather than the initial stage. The crack opening displacement is reduced due to the restraining effect of the CFRP patches, which could lead to an enhanced fatigue life.

Some examples of improvement in fatigue performance due to FRP strengthening are summarised in Table 7.1 for steel plates with various

Table 7.1 Improvement in fatigue performance for steel plates

FRP used (Young's modulus)	Increase in fatigue life	Reference
Steel plate with an initial crack (similar to Figure 7.1(a)(i))		
CFRP plate (165 GPa)	2.2 times	Jones and Civjan (2003)
CFRP plate (174 GPa)	3 times	Colombi et al. (2003)
CFRP plate (174 GPa) with prestress of 632 MPa	5 times	Colombi et al. (2003)
CFRP plate (216 GPa) with prestress of 632 MPa	16 times	Colombi et al. (2003)
CFRP plates (165 GPa)	3.6 times	Zheng et al. (2006)
CFRP plates (320 GPa)	From 5.5 to 6.8 times	Zheng et al. (2006)
CFRP plates (155 GPa)	About 3 times	Täljsten et al. (2009)
CFRP plates (260 GPa)	3.7 times	Täljsten et al. (2009)
CFRP plates (155 GPa) with prestress of 214 MPa	About 8 to10 times	Täljsten et al. (2009)
3 to 5 layers of CFRP sheet (230 GPa)	From 2.2 to 2.7 times	Liu et al. (2009a)
3 to 5 layers of CFRP sheet (552 GPa)	From 4.7 to 8 times	Liu et al. (2009a)
CFRP plates (478 GPa) with various schemes defined in Figure 7.2	From 3.3 to 7.5 times	Wu et al. (2012a)
Steel plate with edge notches as initial cracks (similar to Figure 7.1(a)(ii))		
CFRP plate (165 GPa)	2.7 times	Jones and Civjan (2003)
CFRP plate (205 GPa)	About 1.5 times	Ye et al. (2010)
Steel plate with saw cut as initial crack and stop holes (see Figure 7.1(a)(iii))		
CFRP plate (155 GPa)	Crack propagation rate reduces 40 to 90% depending on the width of CFRP plate	Suzuki (2004)

types of initial cracks and Table 7.2 for I-section beams and welded connections. Tables 7.1 and 7.2 demonstrate that the increased fatigue life because of the FRP strengthening depends on many factors, such as the type of specimen configuration, the type of FRP (carbon or glass, sheet or plate) with or without prestress, the Young's modulus of FRP, the number of FRP layers, and the types of strengthening schemes. Fatigue life also depends on the stress range applied. It should be noted that the same nominal stress range was applied to each of the comparisons with or without FRP.

For example, Suzuki (2004) studied steel plate with saw cut as initial crack and two stop holes (see Figure 7.1(a)(iii)). The ratio of the CFRP plate width (c) to the overall length of the initial crack (a) was taken as 0 (without CFRP), 0.4 (CFRP covering part of the saw cut), and 1.0 (CFRP covering

Table 7.2 Improvement in fatigue performance for I-section beams and welded connections

Configuration	FRP used (Young's modulus)	Improved performance	Reference
I-section steel beams with initial cracks in tension flanges (see Figure 7.1(b))	CFRP plate (144 GPa)	Stable crack growth rate reduces 65%	Tavakkolizadeh and Saadatmanesh (2003)
I-section steel beams with initial cracks in tension flanges (similar to Figure 7.1(b))	4 layers of CFRP sheet (230 GPa)	Fatigue life increases 50% to 4 times	Jiao et al. (2012)
Welded non-load carrying cruciform joints (see Figure 7.1 (c)(i)) without initial cracks	CFRP sheet (240 GPa)	Reduced stress range was found due to CFRP strengthening	Chen et al. (2012)
Welded longitudinal attachment with an initial crack (similar to Figure 7.1(c)(ii))	CFRP plate (188 GPa) up to 5 layers	Fatigue life increases 4 to 10 times	Nakamura et al. (2009)
Welded cross-beam connections (already damaged under fatigue loading); see Figure 7.1(d)(i)	CFRP sheet (230 GPa)	Fatigue life extends twice the original fatigue life	Xiao and Zhao (2012)
Welded aluminium K-joints (see Figure 7.1(d)(ii))	GFRP sheet (30 GPa)	The repaired connections exceeded the fatigue limit of the aluminium welded connections with no known cracks; the repaired connections with 90% of the weld removed satisfied the constant amplitude fatigue limit threshold	Nadauld and Pantelides (2007)

the whole crack). The crack propagation rate of the specimen with c/a ratio of 0.4 was found to be 60% of that without CFRP. The crack propagation rate of the specimen with c/a ratio of 1.0 was found to be less than 10% of that without CFRP. A similar phenomenon was found by Jones and Civjan (2003) and Wu et al. (2012a) through comparing five schemes defined in Figure 7.2.

The influence of Young's modulus of FRP on fatigue life is obvious from Table 7.1. For steel plate with an initial crack (similar to Figure 7.1(a)(i))

Zheng et al. (2006) found that the fatigue life increased 3.6 times if a 165 GPa CFRP plate was used, whereas the fatigue life increased up to 6.8 times if a 320 GPa CFRP plate was used. Liu et al. (2007, 2009a) found that normal-modulus (230 GPa) CFRP sheet strengthening increased fatigue life up to 2.7 times, whereas fatigue life increase reached 8 times if high-modulus (552 GPa) CFRP sheet was used. Wu et al. (2012a) confirmed this phenomenon by using high-modulus (478 GPa) CFRP plate. Prestressed CFRP was found to further enhance the fatigue life improvement. For steel plate with an initial crack (similar to Figure 7.1(a)(i)) Colombi et al. (2003) demonstrated that the increased fatigue life went from three times (without prestress) to five times if prestress of 632 MPa was applied to a 174 GPa CFRP plate. The increased fatigue life increased by 16 times if the same amount of prestress was applied to a 216 GPa CFRP plate. The study by Täljsten et al. (2009) showed that the increased fatigue life went from 3 times (without prestress) to 8 or 10 times (with prestress). Figure 7.3 gives some examples of the influence of Young's modulus of CFRP on fatigue strengthening.

7.4 FATIGUE CRACK PROPAGATION

It is important to understand how fatigue crack propagates in a metallic structure because the fatigue crack propagation rate is directly related to fatigue life. There are several methods to capture the crack propagation in FRP-steel composite systems, e.g., alternating current potential drop

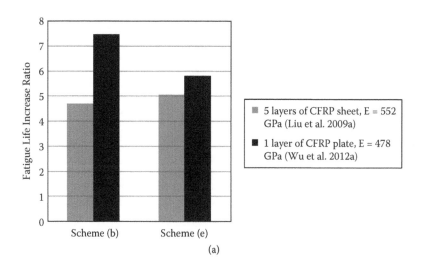

Figure 7.3 Influence of Young's modulus of CFRP and prestress on fatigue strengthening. (a) Comparison of CFRP sheet and CFRP plate strengthening.

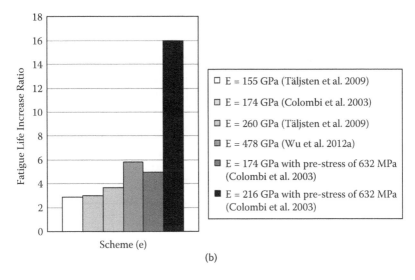

Figure 7.3 (Continued) Influence of Young's modulus of CFRP and prestress on fatigue strengthening. (b) Comparison of CFRP plate strengthening.

(ACPD) method, alternating current field measurement (ACFM) method, crack gauge method, and beach marking method.

The ACPD method is a crack sizing technique that can be used for monitoring and measuring fatigue crack length and depth. It has been widely used to capture fatigue crack propagation in large-scale welded tubular joints (Lie et al. 2005). This method requires probes to be welded on the surface of the specimen. Therefore, it is only suitable for single-sided repair. However, it involves sophisticated skills and subjective interpretation of test results. The ACFM method is an electromagnetic technique for nondestructive testing detection and sizing of surface breaking cracks. This method can be used for double-sided repair since it is able to detect the crack in steel plates under FRP composites. However, the required skill level is very high, and the interpretation of test results could be subjective.

Nakamura et al. (2009) used crack gauges to measure crack length in a CFRP–steel system. However, the crack gauges were only placed outside the CFRP area. No information could be provided regarding the crack propagation within the CFRP-repaired region.

The beach marking method creates visible marks on the fracture surface by varying the applied stress ranges, as illustrated in Figure 7.4. The large stress range is used for crack propagation, whereas the small range is used to create the beach marking because of varied crack propagation rates. Liu et al. (2009a) and Wu et al. (2012a) utilised this method to measure crack propagation of CFRP-strengthened steel plates. Some

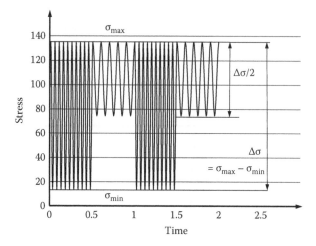

Figure 7.4 Schematic view of loading history to create beach marks.

examples are shown in Figure 7.5. It can be clearly seen that the beach marks are symmetric for the case without CFRP and the case with double-sided repair. The beach marks for the case with single-sided repair (Figure 7.5(b)) clearly show that crack propagates more slowly on the side with CFRP.

Some examples of measured crack propagation against the number of fatigue cycles are presented in Figure 7.6. The fatigue crack grows faster when the crack reaches the edge of the steel plate. It is obvious that the crack propagation rate slows down because of CFRP strengthening, especially those with prestressed CFRP.

7.5 PREDICTION OF FATIGUE LIFE FOR CCT (CENTRE-CRACKED TENSILE) STEEL PLATES STRENGTHENED BY MULTIPLE LAYERS OF CFRP SHEET

7.5.1 Boundary element method approach

7.5.1.1 Boundary element method

The finite element method (FEM) is a commonly used technique for stress analysis because it can model extremely complex configurations and easily determine the response at any desired point of a structure. It has also been used in modelling of composite repairs (Duong and Wang 2007). Similar to the finite element method, the boundary element method (BEM) is another frequently used and well-established

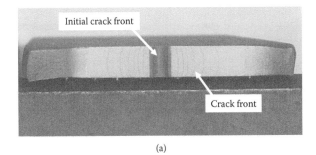

(a)

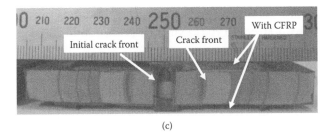

(b)

(c)

Figure 7.5 Examples of observed beach marks. (a) Steel plate without CFRP. (Courtesy of H.B. Liu, Monash University, Australia.) (b) Single-sided repair (strengthening scheme (a) shown in Figure 7.2). (Courtesy of H.B. Liu, Monash University, Australia.) (c) Double-sided repair (strengthening scheme (e) shown in Figure 7.2). (Courtesy of C. Wu, Monash University, Australia.)

numerical technique (Aliabadi and Rooke 1991). It has been widely adopted to analyse composite-repaired metallic structures. Some of the pioneer work was done by Cruse and Besuner (1975), Dowrick et al. (1980), and Ingraffea et al. (1983). Young et al. (1992) and Wen et al. (2002, 2003) used BEM in the analysis of reinforced cracked sheets, in which the patch and cracked sheet were modelled using two-dimensional

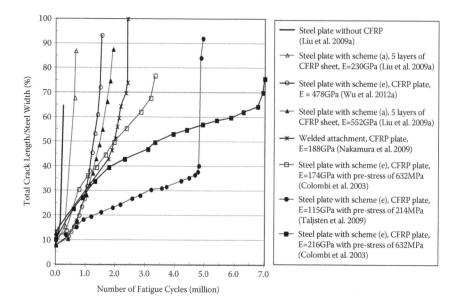

Figure 7.6 Examples of measured crack propagation against number of fatigue cycles.

boundary elements and the adhesive was simulated as shear springs. Three-dimensional fatigue crack growth was simulated using BEM by Mi and Aliabadi (1992, 1994), Mashiri et al. (2000), and Yang et al. (2007).

Both FEM and BEM have been found to provide reasonable results for composite repairs from the literature mentioned above. When used to analyse problems involving crack propagation, BEM may be more efficient (Mashiri et al. 2000). One of the main advantages of BEM is that only a mesh of the surfaces is required, making it easier to use and often less computationally time-consuming. Because this method requires only boundary discretisation, remeshing work for the simulation of the crack growth process is reduced compared to FEM, especially for three-dimensional analysis.

7.5.1.2 BEM model of CCT steel plates strengthened by multiple layers of CFRP sheet

Liu et al. (2009b) carried out a study on fatigue of CCT steel plates strengthened by multiple layers of CFRP sheet. The computer program BEASY (BEASY 2006) was adopted. The composite patch and the cracked steel plate were simulated using surface elements, whereas the adhesive layer was simulated as interface springs to connect the patch and steel plate. The dimensions of specimens simulated are those given in Liu et al. (2009a).

Each specimen was modelled three-dimensionally, and only half of the specimen was meshed due to symmetric configuration. The perspective and cross section views of a double-sided repair model are shown in Figure 7.7(a) and (b). The crack front at each increment is also depicted in Figure 7.7(b), in which AA' represents the initial crack front, BB' represents the crack

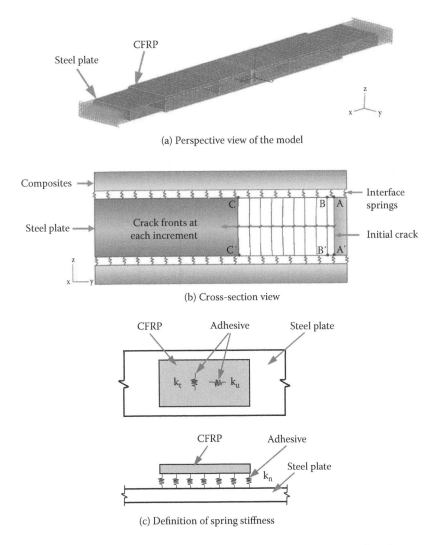

(a) Perspective view of the model

(b) Cross-section view

(c) Definition of spring stiffness

Figure 7.7 BEM model. (From Liu, H.B., Zhao, X.L., and Al-Mahaidi, R., *Composite Structures*, 91(1), 74–83, 2009b.)

front at the first increment, and the direction of crack propagation is from AA' to CC'.

The steel plate and CFRP composites were modelled as continuous plates composed of several zones. Quadrilateral elements, Q38, were placed on all external boundaries and interface surfaces. The central hole in the steel plate was also meshed. The initial 0.1 mm long slot was not meshed because it was defined as an initial crack using BEASY Fracture Wizard. BEASY Fatigue and Crack Growth modules add the crack into the model by simply selecting one crack type from the program's crack library and supplying the number of the initial point and its orientation data.

The equivalent elastic modulus of the composites is needed in the modelling. This is partly because the composites are formed by two types of materials (carbon fibres and structural adhesive), and partly because the strain distribution through the CFRP layers is not uniform. Three or five CFRP sheets were applied to the CCT steel plates using the wet lay-up method (Liu et al. 2009a). The strain distribution across layers of high-modulus CFRP sheet and normal-modulus CFRP sheet is given in Figure 7.8(a) and (b). The equivalent composite concept is illustrated in

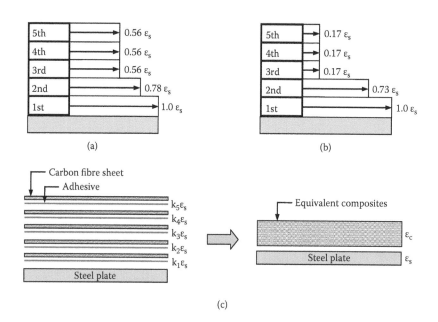

(a)

(b)

(c)

Figure 7.8 Strain distributions (not to scale). (a) Strain distribution across layers of high-modulus CFRP sheet. (b) Strain distribution across layers of normal-modulus CFRP sheet. (c) Illustration of the equivalent composites. (From Liu, H.B. et al., Thin-Walled Structures, 47(10), 1069–1077, 2009c.)

Figure 7.8(c). The equivalent modulus of the composites (E_c) can be derived using the equilibrium condition of forces in the equivalent composites (P_c), fibre (P_f) and adhesive (P_a).

$$P_c = P_f + P_a \tag{7.1}$$

$$P_f = E_f \cdot \left(\sum_{i=1}^{n} k_i \varepsilon_s \right) \cdot t_{CFRP\_sheet} \tag{7.2}$$

$$P_a = E_a \cdot \left(\sum_{i=1}^{n} k_i \varepsilon_s \right) \cdot t_{a\_layer} \tag{7.3}$$

$$P_c = E_c \cdot \varepsilon_c \cdot t_{CFRP} \tag{7.4}$$

$$t_{CFRP} = n \cdot (t_{CFRP\_sheet} + t_{a\_layer}) \tag{7.5}$$

In calculating the equivalent modulus of the composites, the CFRP sheet and adhesive are treated as one material. Assuming there is a perfect bond between the composites and steel plate, the average strain of the composites (ε_c) is equal to the strain of the steel plate (ε_s):

$$\varepsilon_c = \varepsilon_s \tag{7.6}$$

From Equations (7.1) to (7.6), the equivalent modulus of the composites (E_c) can be expressed as

$$E_c = \frac{E_{CFRP\_sheet} \cdot t_{CFRP\_sheet} + E_a \cdot t_{a\_layer}}{n \cdot (t_{CFRP\_sheet} + t_{a\_layer})} \cdot \sum_{i=1}^{n} k_i \tag{7.7}$$

where E_c, E_{CFRP_sheet}, and E_a are Young's moduli of equivalent composites, CFRP fibre, and adhesive, respectively. t_{CFRP} is the thickness of the equivalent composites, t_{CFRP_sheet} is the thickness of one layer of CFRP sheet, t_{a_layer} is the thickness of one layer of adhesive, and n (= 3 or 5) is the number of CFRP layers. The term k_i is the factor used to define the level of strain in the ith layer with respect to the strain in the steel. The values of k_i are summarised as follows, which were derived from the experiments reported in Fawzia (2007) and Liu (2008).

For the strengthening using a normal-modulus CFRP sheet:

$$k_1 = 1.0$$

$k_2 = 0.73$

$k_3 = 0.17$ (7.8)

$k_4 = 0.17$

$k_5 = 0.17$

For the strengthening using a high-modulus CFRP sheet:

$k_1 = 1.0$

$k_2 = 0.78$

$k_3 = 0.56$ (7.9)

$k_4 = 0.56$

$k_5 = 0.56$

The structural adhesive bonding the composites and steel plate was modelled by interface elements connecting the patch and the steel plate in the BEASY program. It is necessary to define proper stiffness values for internal springs of the interface. In the BEASY program spring stiffness is defined in terms of normal and tangential directions in a local coordinate system, as illustrated in Figure 7.7(c). The stiffness value, specified in units of MPa/mm, can be derived using the elastic properties of the structural adhesive and the measured bond line thickness. The formulae for in-plane stiffness (k_t and k_u) and out-of-plane stiffness (k_n) of the adhesive were defined in Liu et al. (2009b).

$$k_t = \frac{G_a}{t_{a\_layer}}$$ (7.10a)

$$k_u = \frac{G_a}{t_{a\_layer}}$$ (7.10b)

$$k_n = \frac{E_a}{t_{a\_layer}}$$ (7.10c)

$$G_a = \frac{E_a}{2(1+v_a)}$$ (7.11)

where G_a and E_a are the shear modulus and Young's modulus of the adhesive, respectively, and v_a is the Poisson's ratio of the adhesive.

Constant amplitude cyclic loading, as in the experimental testing, was used in the modelling. When the models are meshed, the material properties, crack incremental size, and cyclic loading can be specified in BEASY Fracture Wizard. The cracks are able to propagate automatically in BEASY until the complete failure is reached.

7.5.1.3 BEM simulation results

Some simulation results are plotted in Figure 7.9(a) in terms of crack length versus the number of fatigue cycles, together with the corresponding experimental data obtained in Liu et al. (2009a). The predicted fatigue life

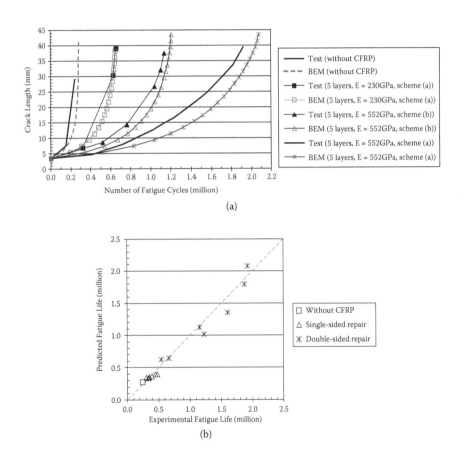

Figure 7.9 Comparison of BEM results and test data. (a) Examples of crack propagation versus fatigue life. (b) Predicted fatigue life versus experimental fatigue life. (Adapted from Liu, H.B. et al., *Composite Structures*, 91(1), 74–83, 2009b.)

from BEM is also compared in Figure 7.9(b) with that obtained from the experimental testing. It seems that the BEM can reasonably simulate the fatigue crack propagation and fatigue life of CCT steel plates strengthened by multiple layers of CFRP sheet.

7.5.2 Fracture mechanics approach

Although many numerical studies based on finite element or boundary element methods have been carried out to predict the stress distribution and crack tip stress intensity factors in a composites-repaired plate, as mentioned in Section 7.5.1, there is a need to develop a relatively simple model to predict fatigue life of the CFRP–steel composite system. Some analytical work was conducted on fatigue of CFRP plate-repaired steel plates (e.g., Colombi 2004, 2005). Very limited theoretical models are available for CFRP sheet repairs, especially when multiple layers of CFRP sheets are used. Liu et al. (2009c) carried out a study to predict the fatigue life CCT steel plates strengthened by multiple layers of CFRP sheet by using the fracture mechanics approach.

7.5.2.1 Fracture mechanics formulae for CCT steel plates

Linear elastic fracture mechanics (LEFM) is widely used to calculate the fatigue crack propagation life of CCT steel plates. The approach is mainly based on the Paris law:

$$da/dN = C \cdot (\Delta K_s^m - \Delta K_{th}^m) \tag{7.12}$$

where N is the number of fatigue cycles, a is the crack length, C and m are empirical material-related constants, ΔK_s is the range of stress intensity factors at the crack tip in a steel plate, and ΔK_{th} is the threshold stress intensity factor below which the fatigue crack does not propagate. The stress intensity factor at the crack tip can be calculated using the following equation (Albrecht and Yamada 1977):

$$K_s = F\sigma_o \sqrt{\pi \cdot a} \tag{7.13}$$

in which F is a crack size-dependent correction factor:

$$F = F_s F_e F_g F_t \tag{7.14}$$

where F_s is the correction factor for surface crack, F_e is the correction factor for crack shape, F_g is the correction factor for stress gradient, and F_t is the correction factor for finite thickness and width of plate.

Factor F modifies the idealised case of a central crack in an infinite plate. In the case of CCT steel plates, the crack in the centre of the steel plate is assumed to be a through-thickness crack that propagates in the width direction under cyclic loading. During the whole crack propagation period nonuniform stress does not exist in the uncracked section. Therefore, the values of F_s, F_e, and F_g are taken as 1.0.

$$F_s = F_e = F_g = 1.0 \tag{7.15}$$

As the steel plates have limited width, the correction factor F_t can be simplified as (Albrecht and Yamada 1977)

$$F_t = \sqrt{\sec\left(\frac{\pi \cdot a}{W_s}\right)} \tag{7.16}$$

The fatigue life ($N_{FM,s}$) can be predicted by integrating Equation (7.12) as

$$N_{FM,s} = \int_{a_i}^{a_f} \frac{1}{C \cdot (\Delta K_s^m - \Delta K_{th}^m)} da \tag{7.17}$$

$$\Delta K_s = K_{s,max} - K_{s,min} \tag{7.18}$$

where a_i and a_f are the initial and final sizes of the crack, and $K_{s,max}$ and $K_{s,min}$ are the maximum and minimum stress intensity factors.

When CFRP sheet is applied, modifications are necessary to the stress value (σ_o) and stress intensity factor range ΔK_s, as explained in Sections 7.5.2.2 and 7.5.2.3.

7.5.2.2 Average stress in steel plate with CFRP sheet

For multiple CFRP sheet-strengthened CCT steel plates, the force equilibrium in the composite-steel system can be illustrated in Figure 7.10(a) for single-sided repair and Figure 7.10(c) for double-sided repair. Geometry and notation at the cracked section are shown in Figure 7.10(b) and (d).

By applying the equilibrium conditions in the CFRP–steel systems,

$$P = P_s + P_c \quad \text{for single-sided repair} \tag{7.19a}$$

$$P = P_s + 2P_c \quad \text{for double-sided repair} \tag{7.19b}$$

$$P = \sigma_s \cdot W_c \cdot t_s \tag{7.20}$$

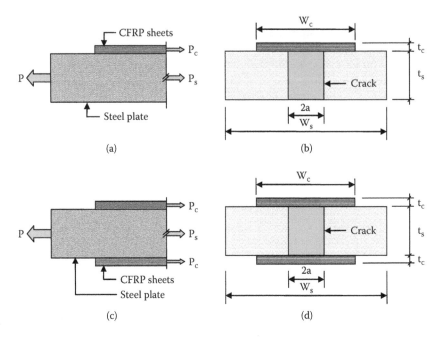

Figure 7.10 Equilibrium conditions. (a) Force equilibrium for single-sided repair. (b) Geometry and notation at the cracked section of single-sided repair. (c) Force equilibrium for double-sided repair. (d) Geometry and notation at the cracked section of double-sided repair. (Adapted from Liu, H.B. et al., *Thin-Walled Structures*, 47(10), 1069–1077, 2009c.)

$$P_s = E_s \cdot \varepsilon_s \cdot (W_s - 2a) \cdot t_s \tag{7.21}$$

$$P_c = E_c \cdot \varepsilon_c \cdot W_c \cdot t_c \tag{7.22}$$

where P is the external force, F_s and F_c are the forces carried by the steel plate and composites, respectively, σ_0 is the nominal stress in the steel plate, W_s is the width of the steel plate, W_c is the width of the CFRP sheet, 2a is the total crack length defined in Figure 7.10, E_s is the Young's modulus of steel plate, and E_c is the equivalent modulus of the composites defined in Equation (7.7). The average strain of the composites (ε_c) is assumed to be equal to the strain of the steel plate (ε_s), as mentioned in Section 7.5.2.1.

From the above-mentioned equations, the strain at the cracked section of the reinforced plates can be expressed as

$$\varepsilon_{s,SG} = \frac{\sigma_0 \cdot W_s \cdot t_s}{E_s \cdot (W_s - 2a) \cdot t_s + E_c \cdot W_c \cdot t_c} \tag{7.23a}$$

$$\varepsilon_{s,DB} = \frac{\sigma_o \cdot W_s \cdot t_s}{E_s \cdot (W_s - 2a) \cdot t_s + 2E_c \cdot W_c \cdot t_c} \qquad (7.23b)$$

where the subscript SG stands for single-sided repair and DB for double-sided repair.

The force carried by the steel plate (P_s) can also be expressed in terms of average stress (σ_s) over the steel section.

$$P_s = \sigma_s \cdot W_s \cdot t_s \qquad (7.24)$$

Substituting Equation (7.24) into Equation (7.21) leads to

$$\sigma_s = E_s \cdot \varepsilon_s \cdot \left(\frac{W_s - 2a}{W_s} \right) \qquad (7.25)$$

Substituting Equation (7.23) into Equation (7.25) leads to

$$\sigma_{s,SG} = \frac{E_s t_s}{E_s t_s + E_c \left(\dfrac{W_c}{W_s - 2a} \right) t_c} \sigma_o \qquad (7.26a)$$

$$\sigma_{s,DB} = \frac{E_s t_s}{E_s t_s + 2E_c \left(\dfrac{W_c}{W_s - 2a} \right) t_c} \sigma_o \qquad (7.26b)$$

where $\sigma_{s,SG}$ and $\sigma_{s,DB}$ are the average stress at the nominal cross section of the steel plate for single-sided repair and double-sided repair, respectively.

7.5.2.3 Effective stress intensity factor in steel plate with CFRP sheet

The single-sided repair is less effective to extend fatigue life of CCT plates when compared with double-sided repair. This is because that crack propagates much faster on the unpatched side (i.e., the critical side to reach complete failure), as shown in Figure 7.5(b). The out-of-plane bending also makes the situation worse, as shown by Ratwani (1979), that the bending effect on the stress intensity factor increases as the crack propagates. However, the crack in the steel plate of single-sided repair could still experience crack closure effect, even though the effect is weakened by the bending moment. It was assumed by Liu et al. (2009c) that the effect of crack closure and the effect of bending moment are counteracted in single-sided repairs, since it is very difficult to quantify the crack closure effect under bending and tension loads. Hence, the same stress intensity factor expression used for CCT plates is adopted for single-sided repair.

$$\Delta K_{eff,SG} = \Delta K_s \tag{7.27}$$

For double-sided repair, it can be assumed that there are two identical reinforcements bonded on each side of the plate. The crack propagation beach mark shown in Figure 7.5(c) supports this assumption. This symmetric arrangement ensures that in-plane loads produce no out-of-plane bending over the repaired zone. The through-thickness crack is assumed to propagate perpendicularly to the plate surface, and the stress intensity factors at the crack tip are uniform. Under this condition the cracks ought to experience the crack closure—a concept first introduced by Elber (1970); i.e., the faces of fatigue cracks in metallic structures can make contact even though the specimens remain in tension.

The premature closure of the crack leads to a retardation of the fatigue crack propagation. Elber therefore proposed that the crack tip did not act as a stress concentrator until the theoretical stress intensity factor reached a certain value, at which the crack faces are fully opened. The stress intensity factor given in Equation (7.13) needs to be replaced by a so-called effective stress intensity factor. Wang (2002) developed a simple expression for such an effective stress intensity factor as

$$\Delta K_{eff,DB} = \frac{\Delta K_s}{1-R} \cdot \left(1 - \frac{\sigma_{op}}{\sigma_{max}}\right) \tag{7.28}$$

$$R = \frac{\sigma_{min}}{\sigma_{max}} \tag{7.29}$$

where ΔK_s is given in Equations (7.13) and (7.18) in Section 7.5.2.1, with σ_o replaced by σ_s given in Equation (7.26). σ_{min} and σ_{max} are the minimum and maximum values of the applied cyclic stress, respectively, and σ_{op} is the crack opening stress, which can be obtained by

$$\frac{\sigma_{op}}{\sigma_{max}} = \begin{cases} A_0 & \text{for } R < 0 \\ A_0 + (2 - 3A_0) \cdot R^2 + (2A_0 - 1) \cdot R^3 & \text{for } R > 0 \end{cases} \tag{7.30}$$

where

$$A_0 = 0.825 - 0.34\alpha + 0.05\alpha^2 \tag{7.31}$$

$$\alpha = \frac{1 + 0.64\left(\sqrt{\dfrac{\omega}{t_s}} + 2\left(\dfrac{\omega}{t_s}\right)^2\right)}{1 - 2v + 0.54\left(\sqrt{\dfrac{\omega}{t_s}} + 2\left(\dfrac{\omega}{t_s}\right)^2\right)} \tag{7.32}$$

$$\omega = \frac{\pi}{8}\left(\frac{K_s}{\sigma_{y,s}}\right)^2 \tag{7.33}$$

in which α is the plastic constraint factor, ω denotes the plastic zone size, $\sigma_{y,s}$ is the yield stress of steel plate, and v is the Poisson ratio of steel.

7.5.2.4 Fatigue life of CCT steel plates strengthened by multiple layers of CFRP sheet

The predicted fatigue life of CCT plates strengthened by multiple layers of CFRP sheet can be calculated by the following integration:

$$N_{FM,SG} = \int_{a_i}^{a_f} \frac{1}{C \cdot (\Delta K_{eff,SG}^m - \Delta K_{th}^m)}\, da \tag{7.34a}$$

$$N_{FM,DB} = \int_{a_i}^{a_f} \frac{1}{C \cdot (\Delta K_{eff,DB}^m - \Delta K_{th}^m)}\, da \tag{7.34b}$$

where the subscript FM stands for fracture mechanics, SG for single-sided repair, and DB for double-sided repair. $\Delta K_{eff,SG}$ is given in Equation (7.27), and $\Delta K_{eff,DB}$ is given in Equation (7.28).

Examples of predicted crack propagation against the number of fatigue cycles are plotted in Figure 7.11(a), together with experimental data reported in Liu et al. (2009a). The values for C, m, and ΔK_{th} are taken as those for the mean design curve for steel in the Japan Society of Steel Construction (JSSC) (1995) fatigue design recommendations. The predicted fatigue life from the fracture mechanics approach is also compared in Figure 7.11(b) with that obtained from the experimental testing. It seems that the fracture mechanics analysis can reasonably simulate the fatigue crack propagation and fatigue life of CCT steel plates strengthened by multiple layers of CFRP sheet.

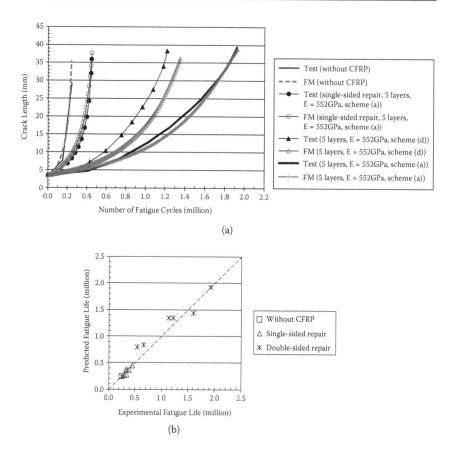

Figure 7.11 Comparison of the fracture mechanics results and experimental data. (a) Examples of crack propagation versus fatigue life. (b) Predicted fatigue life versus experimental fatigue life. (Adapted from Liu, H.B. et al., *Thin-Walled Structures*, 47(10), 1069–1077, 2009c.)

7.6 STRESS INTENSITY FACTOR FOR CCT STEEL PLATES STRENGTHENED BY CFRP

7.6.1 Existing approaches

A few analytical methods have been developed to predict the fatigue life of FRP laminate-repaired aluminium panels (Baker et al. 2002, Duong and Wang 2004, Hosseini and Mohammadi 2007, Seo and Lee 2002). They were developed for aluminium panels repaired by using glass-epoxy, boron-epoxy, and graphite-epoxy systems. The results may not be directly applied to CFRP-repaired steel plates because of different failure modes and different material properties. Generally, there are three approaches for obtaining the stress intensity factor (SIF) of CFRP-reinforced steel plates: analytical, numerical, and empirical (Shen and Hou 2011).

For analytical solutions, the stress field at the crack tip becomes very complicated when considering CFRP and adhesive effects (Colombi 2004, 2005). In Colombi (2005) the fatigue crack propagation rate was postulated to be a function of the effective strain energy density factor range. Fatigue crack growth data showed that the standard crack growth retardation model cannot be used to evaluate the minimum effective stress. Hence, an ad hoc plasticity model was introduced and validated using experimental results. The proposed technique is an extension of the well-known Newman's model. The bridging effect due to the reinforcing strips is analytically modelled in order to estimate the reduction of crack opening displacement, and finally the magnification of the crack growth retardation.

Numerical simulation has been widely adopted in aerospace engineering to investigate the SIFs of cracked aluminium components repaired by adhesively bonded composites (Gu et al. 2010, Baker et al. 2002, Lee and Lee 2004, Sabelkin et al. 2007). However, interface damage and deterioration of the bond during fatigue loading are still unsolved problems (Naboulsi and Mall 1996, Colombi et al. 2003, Buyukozturk et al. 2004).

SIF can be empirically obtained through experimental observations of crack propagation against the number of fatigue cycles, e.g., using the James-Anderson method developed in the 1969 (James and Anderson 1969). Shen and Hou (2011) adopted the James-Anderson method and proposed a SIF formula for CCT aluminium plates strengthened with single-sided CFRP composite. The experimental SIFs were derived from the crack growth curves (crack length versus fatigue cycles) of the specimens. A modifying factor was introduced as a linear function of crack width-to-plate width ratio. However, the effect of CFRP mechanical properties was not incorporated in their SIF expression, limiting its applicability to different CFRP systems.

Wu et al. (2012b) attempted to use the empirical approach to derive SIF formulae for CCT steel plates strengthened by CFRP. It considered the effective stress in the steel plate due to CFRP and the influence of crack length and CFRP bond width on SIF. The aim is to present SIF formulae that can be expressed explicitly in terms of crack length (a), geometric dimensions (W_s, W_c, t_s, t_c), and material properties (E_s and E_c). This approach is described in detail in this section.

It should be noted that the following CFRP and adhesive were used in the experimental program on which the derivations presented in this section were based: CFRP sheet with measured elastic moduli of 230 GPa and 552 GPa, CFRP plate with measured elastic modulus of 479 GPa, and adhesive Araldite 420 (Liu et al. 2009a, Wu et al. 2012a).

7.6.2 Stress intensity factor for CCT steel plates without CFRP

A classic solution is available (Tada et al. 1985, Anderson 2005) for the SIF range (ΔK_s) of CCT steel plates without CFRP strengthening. It can be summarised as follows:

$$\Delta K_s = F\Delta\sigma_o\sqrt{\pi \cdot a} \tag{7.35}$$

$$\Delta\sigma_o = \sigma_{o,max} - \sigma_{o,min} \tag{7.36}$$

$$F = F_s F_e F_g F_h F_t \tag{7.37}$$

$$F_s = F_e = F_g = F_h = 1.0 \tag{7.38}$$

where F_s is the correction factor for surface crack, F_e is the correction factor for crack shape, F_g is the correction factor for stress gradient, F_h is the correction factor for eccentricity of crack against the central axis of the plate, and F_t is the correction factor for finite thickness and width of plate.

$$F_t = (1-0.025\lambda^2 +0.06\lambda^4)\sqrt{\sec\left(\frac{\pi \cdot \lambda}{2}\right)} \tag{7.39}$$

$$\lambda = \frac{2a}{W_s} \tag{7.40}$$

Equation (7.35) can be simplified to

$$\Delta K_s = F_t\Delta\sigma_o\sqrt{\pi \cdot a} \tag{7.41}$$

7.6.3 Influence on stresses in steel plate due to CFRP

Wu et al. (2012b) introduced two factors to consider the effect of CFRP on stress ranges and the effect of crack length and CFRP width. Only one layer of CFRP plate was applied. The influence on stresses in steel plate due to CFRP is discussed in this section, whereas the influence of crack length and CFRP width will be described in Section 7.6.4.

Similar to the approach described in Section 7.5.2.2, from force equilibrium conditions, the same expressions as those in Equation (7.26) could be obtained. The only difference is that the term ($W_s - 2a$) is replaced by W_s

since the influence of the crack length (a) will be considered separately later (see Section 7.6.4). Hence,

$$\sigma_{s,SG} = \frac{E_s t_s}{E_s t_s + E_c \left(\dfrac{W_c}{W_s}\right) t_c} \sigma_o \tag{7.42a}$$

$$\sigma_{s,DB} = \frac{E_s t_s}{E_s t_s + 2E_c \left(\dfrac{W_c}{W_s}\right) t_c} \sigma_o \tag{7.42b}$$

where $\sigma_{s,SG}$ and $\sigma_{s,DB}$ are the average stress at the nominal cross section of the steel plate for single-sided repair and double-sided repair, respectively. W_s is the steel plate width, W_c is the CFRP plate width, and E_s and t_s are the Young's modulus and thickness of the steel plate.

t_c is the thickness of the CFRP composite (including CFRP and adhesive):

$$t_c = t_{CFRP\_plate} + t_a \tag{7.43}$$

where t_{CFRP_plate} is the thickness of one layer of CFRP plate and t_a is the thickness of the adhesive between the steel plate and the first layer of CFRP.

E_c can be derived in a manner similar to that for the CFRP sheet expressed in Equation (7.7), except that only one layer of CFRP needs to be considered here, i.e.,

$$E_c = \frac{E_{CFRP} \cdot t_{CFRP\_plate} + E_a \cdot t_a}{t_{CFRP\_plate} + t_a} \tag{7.44}$$

Equation (7.42) can be rewritten as

$$\sigma_{s,SG} = \frac{1}{1 + \dfrac{E_c W_c t_c}{E_s W_s t_s}} \sigma_o = F_{p,SG} \cdot \sigma_o \tag{7.45a}$$

$$\sigma_{s,DB} = \frac{1}{1 + 2\dfrac{E_c W_c t_c}{E_s W_s t_s}} \sigma_o = F_{p,DB} \cdot \sigma_o \tag{7.45b}$$

$$F_{p,SG} = \frac{1}{1 + \beta_{SG}} \tag{7.46a}$$

$$F_{p,DB} = \frac{1}{1 + \beta_{DB}} \tag{7.46b}$$

where

$$\beta_{SG} = \frac{E_c W_c t_c}{E_s W_s t_s} \tag{7.47a}$$

$$\beta_{DB} = 2\frac{E_c W_c t_c}{E_s W_s t_s} \tag{7.47b}$$

7.6.4 Influence of crack length and CFRP bond width on SIF

The SIF values also depend on the crack length and CFRP bond width. A reduction factor (F_w) was proposed by Wu et al. (2012b) to consider the influence of crack length and CFRP bond width. It can be expressed as a function of ξ defined in Equation (7.48).

$$\xi = \lambda \cdot \left(\frac{W_c}{W_s}\right) = \left(\frac{2a}{W_s}\right) \cdot \left(\frac{W_c}{W_s}\right) \tag{7.48}$$

The first term λ of $2a/W_s$ is commonly used in fracture mechanics theory, as in Equation (7.40). It refers to the percentage of crack propagation along a CCT steel plate. When $2a/W_s$ becomes 1.0, it means the crack grows fully across the steel plate. The second term W_c/W_s refers to the amount of CFRP applied on the CCT steel plate. When W_c/W_s equals 1.0, it means that the steel plate is fully covered by CFRP (e.g., scheme (a) defined in Figure 7.2). When W_c/W_s is less than 1.0, it refers to other schemes (e.g., schemes (b) to (e)).

The reduction factor F_w (ξ) will be determined empirically based on experimental results in the literature. The ratio of the SIF with CFRP ($\Delta K_{s,CFRP}$) to that without CFRP (ΔK_s) can be written as

$$\frac{\Delta K_{s,CFRP}}{\Delta K_s} = F_p \cdot F_w(\xi) \tag{7.49}$$

Hence,

$$F_w(\xi) = \frac{\Delta K_{s,CFRP}}{F_p \Delta K_s} \tag{7.50}$$

where F_p is given in Equation (7.46) and ΔK_s is given in Equation (7.41).
From the classical Paris Power law,

$$\frac{da}{dN} = C(\Delta K)^m \tag{7.51}$$

The stress intensity factor for CCT plates with CFRP ($\Delta K_{s,CFRP}$) at crack length $\hat{a}$ can be expressed as

$$\left(\Delta K_{s,CFRP}\right)_{\hat{a}} = \left\{\frac{1}{C}\left(\frac{da}{dN}\right)_{\hat{a}}\right\}^{1/m} \tag{7.52}$$

The secant method given in ASTM E647 (2001) was adopted to calculate the crack propagation rate (da/dN) at crack length $\hat{a}$.

$$\left(\frac{da}{dN}\right)_{\hat{a}} = \frac{a_{i+1} - a_i}{N_{i+1} - N_i} \tag{7.53}$$

where a_i and a_{i+1} are crack length at ith and (i + 1)th steps, and N_i and N_{i+1} are the corresponding number of cycles. This is an average crack growth rate over the crack length increment $(a_{i+1} - a_i)$, corresponding to a crack length $\hat{a}$ of $(a_{i+1} + a_i)/2$.

The reduction factor F_w can be calculated using Equations (7.41), (7.46), (7.50), (7.52), and (7.53) based on the test results in Liu et al. (2009a), Täljsten et al. (2009), Zheng et al. (2006), and Wu et al. (2012a). The calculated F_w values are plotted in Figure 7.12 against the parameter ξ of $(2a/W_s)(W_c/W_s)$. For simplicity, three sets of formulae were proposed (Wu et al. 2012b) to cover the scatter of the data: the upper bound, mean, and lower bound formulae.

For single-sided repair the expression can be approximated as

$$F_{w,SG,upper} = \begin{cases} 1.0 & \text{for } 0.1 \le \xi < 0.6 \\ 1.3 - 0.5\xi & \text{for } 0.6 \le \xi < 0.9 \end{cases} \tag{7.54a}$$

$$F_{w,SG,mean} = \begin{cases} 0.9 & \text{for } 0.1 \le \xi < 0.5 \\ 1.15 - 0.5\xi & \text{for } 0.5 \le \xi < 0.9 \end{cases} \tag{7.54b}$$

$$F_{w,SG,lower} = \begin{cases} 0.8 & \text{for } 0.1 \le \xi < 0.4 \\ 1.0 - 0.5\xi & \text{for } 0.4 \le \xi < 0.9 \end{cases} \tag{7.54c}$$

For double-sided repair the expression can be approximated as

$$F_{w,DB,upper} = \begin{cases} 0.8 & \text{for } 0.1 \le \xi < 0.4 \\ 1.04 - 0.6\xi & \text{for } 0.4 \le \xi < 0.9 \end{cases} \tag{7.55a}$$

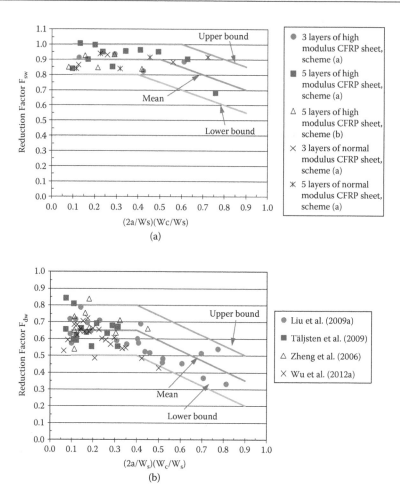

Figure 7.12 Bond width reduction factor for (a) single-side reinforced CCT steel plates (test results from Liu, H. B. et al. 2009) and (b) double-side reinforced CCT steel plates. (Adapted from Wu, C. et al., *International Journal of Structural Stability and Dynamics*, 2012, in press.)

$$F_{w,DB,mean} = \begin{cases} 0.65 & \text{for } 0.1 \le \xi < 0.4 \\ 0.89 - 0.6\xi & \text{for } 0.4 \le \xi < 0.9 \end{cases} \qquad (7.55b)$$

$$F_{w,DB,lower} = \begin{cases} 0.50 & \text{for } 0.1 \le \xi < 0.4 \\ 0.74 - 0.6\xi & \text{for } 0.4 \le \xi < 0.9 \end{cases} \qquad (7.55c)$$

It can be seen from Figure 7.12 and Equations (7.54) and (7.55) that when the crack propagates further (i.e., greater $2a/W_s$ value) or a wider CFRP is applied (i.e., larger W_c/W_s ratio), the value of SIF reduction factor F_w becomes smaller (meaning more reduction in SIF). It is obvious from Figure 7.12 that more reduction in SIF is achieved by applying CFRP on both sides of the steel plates. It seems that the benefit of CFRP reinforcement increases (more reduction of SIF) as the factor ξ exceeds a certain value, as indicated by the point beyond which the slope of the curves in Figure 7.12 changes. This observation can be explained from two aspects: For a specific specimen (i.e., a fixed value of W_c/W_s), when the crack grows further (i.e., greater $2a/W_s$), the steel section is reduced and CFRP will carry a larger portion of fatigue loading. On the other hand, for the same crack length (i.e., as fixed $2a/W_s$), the SIF will obviously reduce more if more CFRP is applied (i.e., larger W_c/W_s).

7.6.5 SIF for CCT steel plates strengthened by CFRP

From Equations (7.41) and (7.49) the SIF for CCT steel plates strengthened by CFRP can be expressed as

$$\Delta K_{s,CFRP} = F_p F_w \Delta K_s = F_t F_p F_w \Delta\sigma_o \sqrt{\pi a} \tag{7.56}$$

where F_t is given in Equation (7.39), F_p is given in Equation (7.46a) for single-sided repair and in Equation (7.46b) for double-sided repair, and F_w is given in Equation (7.54) for single-sided repair and in Equation (7.55) for double-sided repair.

Equation (7.56) can be expanded in terms of crack length (a), geometric dimensions (W_s, W_c, t_s, t_c) and material properties (E_s and E_c).

For single-sided repair:

$\Delta K_{s,CFRP,SG}$

$$= \left(1 - 0.025\left(\frac{2a}{W_s}\right)^2 + 0.06\left(\frac{2a}{W_s}\right)^4\right)\sqrt{\sec\left(\frac{\pi \cdot a}{W_s}\right)}\left(\frac{1}{1 + \dfrac{E_c W_c t_c}{E_s W_s t_s}}\right) F_{w,SG}\Delta\sigma_o\sqrt{\pi a} \tag{7.57a}$$

$$F_{w,SG,upper} = \begin{cases} 1.0 & \text{for } 0.1 \leq \left(\dfrac{2a}{W_s}\right)\left(\dfrac{W_c}{W_s}\right) < 0.6 \\[3mm] 1.3 - 0.5 \cdot \left(\dfrac{2a}{W_s}\right)\left(\dfrac{W_c}{W_s}\right) & \text{for } 0.6 \leq \left(\dfrac{2a}{W_s}\right)\left(\dfrac{W_c}{W_s}\right) < 0.9 \end{cases} \tag{7.57b}$$

$$F_{w,SG,mean} = \begin{cases} 0.9 & \text{for } 0.1 \le \left(\dfrac{2a}{W_s}\right)\left(\dfrac{W_c}{W_s}\right) < 0.5 \\[3ex] 1.15 - 0.5 \cdot \left(\dfrac{2a}{W_s}\right)\left(\dfrac{W_c}{W_s}\right) & \text{for } 0.5 \le \left(\dfrac{2a}{W_s}\right)\left(\dfrac{W_c}{W_s}\right) < 0.9 \end{cases} \tag{7.57c}$$

$$F_{w,SG,lower} = \begin{cases} 0.8 & \text{for } 0.1 \le \left(\dfrac{2a}{W_s}\right)\left(\dfrac{W_c}{W_s}\right) < 0.4 \\[3ex] 1.0 - 0.5 \cdot \left(\dfrac{2a}{W_s}\right)\left(\dfrac{W_c}{W_s}\right) & \text{for } 0.4 \le \left(\dfrac{2a}{W_s}\right)\left(\dfrac{W_c}{W_s}\right) < 0.9 \end{cases} \tag{7.57d}$$

For double-sided repair:

$$\Delta K_{s,CFRP,DB}$$

$$= \left(1 - 0.025\left(\frac{2a}{W_s}\right)^2 + 0.06\left(\frac{2a}{W_s}\right)^4\right)\sqrt{\sec\left(\frac{\pi \cdot a}{W_s}\right)}\left(\frac{1}{1 + 2\dfrac{E_c W_c t_c}{E_s W_s t_s}}\right)F_{w,DB}\Delta\sigma_o\sqrt{\pi a} \tag{7.58a}$$

$$F_{w,DB,upper} = \begin{cases} 0.8 & \text{for } 0.1 \le \left(\dfrac{2a}{W_s}\right)\left(\dfrac{W_c}{W_s}\right) < 0.4 \\[3ex] 1.04 - 0.6 \cdot \left(\dfrac{2a}{W_s}\right)\left(\dfrac{W_c}{W_s}\right) & \text{for } 0.4 \le \left(\dfrac{2a}{W_s}\right)\left(\dfrac{W_c}{W_s}\right) < 0.9 \end{cases} \tag{7.58b}$$

$$F_{w,DB,mean} = \begin{cases} 0.65 & \text{for } 0.1 \le \left(\dfrac{2a}{W_s}\right)\left(\dfrac{W_c}{W_s}\right) < 0.4 \\[3ex] 0.89 - 0.6 \cdot \left(\dfrac{2a}{W_s}\right)\left(\dfrac{W_c}{W_s}\right) & \text{for } 0.4 \le \left(\dfrac{2a}{W_s}\right)\left(\dfrac{W_c}{W_s}\right) < 0.9 \end{cases} \tag{7.58c}$$

$$F_{w,DB,lower} = \begin{cases} 0.50 & \text{for } 0.1 \le \left(\dfrac{2a}{W_s}\right)\left(\dfrac{W_c}{W_s}\right) < 0.4 \\[3ex] 0.74 - 0.6 \cdot \left(\dfrac{2a}{W_s}\right)\left(\dfrac{W_c}{W_s}\right) & \text{for } 0.4 \le \left(\dfrac{2a}{W_s}\right)\left(\dfrac{W_c}{W_s}\right) < 0.9 \end{cases} \tag{7.58d}$$

7.6.6 Influence of key parameters on SIF reduction due to CFRP strengthening

It can be seen from Equations (7.57) and (7.58) that the key parameters influencing SIF are E_c/E_s, W_c/W_s, t_c/t_s, and $2a/W_s$. The ratio of the SIF with CFRP to that without CFRP gives an indication of the reduction in SIF. From Equation (7.56), the ratio can be written as

$$\frac{\Delta K_{s,CFRP}}{\Delta K_s} = F_p \cdot F_w \qquad (7.59)$$

which is a function of E_c/E_s, W_c/W_s, t_c/t_s, and $2a/W_s$. These parameters could practically vary in the following ranges:

$$0.5 \leq \frac{E_c}{E_s} \leq 3 \qquad (7.60a)$$

$$0.56 \leq \frac{W_c}{W_s} \leq 1 \qquad (7.60b)$$

$$0.1 \leq \frac{t_c}{t_s} \leq 0.8 \qquad (7.60c)$$

$$0 \leq \frac{2a}{W_s} \leq 1 \qquad (7.60d)$$

Eight different cases are chosen in this section as examples to show the influence of these key parameters on the SIF ratio defined in Equation (7.59). The examples cover certain combinations of the four parameters. The mean value of F_w from Equation (7.57c) for single-sided repair (SG) and Equation (7.58c) for double-sided repair (DB) is adopted as a typical example.

The eight cases are described below. Equation (7.59) for each case is listed in Table 7.3.

Case 1: Choose strengthening scheme (a) with normal-modulus CFRP to show the influence of t_c/t_s as crack propagates (i.e., $2a/W_s$ increases). In this case, W_c/W_s can be taken as 1.0 because of strengthening scheme (a) shown in Figure 7.2(a). E_c/E_s is taken as 1.0 as a representative case for normal-modulus CFRP.

Case 2: Choose strengthening scheme (a) with high-modulus CFRP to show the influence of t_c/t_s as crack propagates (i.e., $2a/W_s$ increases). In this case, W_c/W_s can be taken as 1.0 because of strengthening

Table 7.3 SIF ratios for eight examples

Case no.	$\Delta K_{s,CFRP,SG}/\Delta K_s$		$\Delta K_{s,CFRP,DB}/\Delta K_s$	
1	$\dfrac{0.9}{1+t_c/t_s}$ for $0.1 \leq 2a/W_s < 0.5$	$\dfrac{1.15-0.5\left(\dfrac{2a}{W_s}\right)}{1+t_c/t_s}$ for $0.5 \leq 2a/W_s < 0.9$	$\dfrac{0.65}{1+2(t_c/t_s)}$ for $0.1 \leq 2a/W_s < 0.4$	$\dfrac{0.89-0.6\left(\dfrac{2a}{W_s}\right)}{1+2(t_c/t_s)}$ for $0.4 \leq 2a/W_s < 0.9$
2	$\dfrac{0.9}{1+2.5(t_c/t_s)}$ for $0.1 \leq 2a/W_s < 0.5$	$\dfrac{1.15-0.5\left(\dfrac{2a}{W_s}\right)}{1+2.5(t_c/t_s)}$ for $0.5 \leq 2a/W_s < 0.9$	$\dfrac{0.65}{1+5(t_c/t_s)}$ for $0.1 \leq 2a/W_s < 0.4$	$\dfrac{0.89-0.6\left(\dfrac{2a}{W_s}\right)}{1+5(t_c/t_s)}$ for $0.4 \leq 2a/W_s < 0.9$
3	$\dfrac{0.9}{1+0.6(t_c/t_s)}$ for $0.17 \leq 2a/W_s < 0.83$	$\dfrac{1.15-0.3\left(\dfrac{2a}{W_s}\right)}{1+0.6(t_c/t_s)}$ for $0.83 \leq 2a/W_s < 1.0$	$\dfrac{0.65}{1+1.2(t_c/t_s)}$ for $0.17 \leq 2a/W_s < 0.83$	$\dfrac{0.89-0.36\left(\dfrac{2a}{W_s}\right)}{1+1.2(t_c/t_s)}$ for $0.83 \leq 2a/W_s < 1.0$
4	$\dfrac{0.9}{1+1.5(t_c/t_s)}$ for $0.17 \leq 2a/W_s < 0.83$	$\dfrac{1.15-0.3\left(\dfrac{2a}{W_s}\right)}{1+1.5(t_c/t_s)}$ for $0.83 \leq 2a/W_s < 1.0$	$\dfrac{0.65}{1+3(t_c/t_s)}$ for $0.17 \leq 2a/W_s < 0.83$	$\dfrac{0.89-0.36\left(\dfrac{2a}{W_s}\right)}{1+3(t_c/t_s)}$ for $0.83 \leq 2a/W_s < 1.0$

5	$\dfrac{0.9}{1+0.2(E_c/E_s)}$ for $0.1 \leq 2a/W_s < 0.5$	$\dfrac{1.15-0.5\left(\dfrac{2a}{W_s}\right)}{1+0.2(E_c/E_s)}$ for $0.5 \leq 2a/W_s < 0.9$	$\dfrac{0.65}{1+0.4(E_c/E_s)}$ for $0.1 \leq 2a/W_s < 0.4$	$\dfrac{0.89-0.6\left(\dfrac{2a}{W_s}\right)}{1+0.4(E_c/E_s)}$ for $0.4 \leq 2a/W_s < 0.9$
6	$\dfrac{0.9}{1+0.5(E_c/E_s)}$ for $0.1 \leq 2a/W_s < 0.5$	$\dfrac{1.15-0.5\left(\dfrac{2a}{W_s}\right)}{1+0.5(E_c/E_s)}$ for $0.5 \leq 2a/W_s < 0.9$	$\dfrac{0.65}{1+E_c/E_s}$ for $0.1 \leq 2a/W_s < 0.4$	$\dfrac{0.89-0.6\left(\dfrac{2a}{W_s}\right)}{1+E_c/E_s}$ for $0.4 \leq 2a/W_s < 0.9$
7	$\dfrac{0.9}{1+0.12(E_c/E_s)}$ for $0.17 \leq 2a/W_s < 0.83$	$\dfrac{1.15-0.3\left(\dfrac{2a}{W_s}\right)}{1+0.12(E_c/E_s)}$ for $0.83 \leq 2a/W_s < 1.0$	$\dfrac{0.65}{1+0.24(E_c/E_s)}$ for $0.17 \leq 2a/W_s < 0.83$	$\dfrac{0.89-0.36\left(\dfrac{2a}{W_s}\right)}{1+0.24(E_c/E_s)}$ for $0.83 \leq 2a/W_s < 1.0$
8	$\dfrac{0.9}{1+0.3(E_c/E_s)}$ for $0.17 \leq 2a/W_s < 0.83$	$\dfrac{1.15-0.3\left(\dfrac{2a}{W_s}\right)}{1+0.3(E_c/E_s)}$ for $0.83 \leq 2a/W_s < 1.0$	$\dfrac{0.65}{1+0.6(E_c/E_s)}$ for $0.17 \leq 2a/W_s < 0.83$	$\dfrac{0.89-0.36\left(\dfrac{2a}{W_s}\right)}{1+0.6(E_c/E_s)}$ for $0.83 \leq 2a/W_s < 1.0$

scheme (a). E_c/E_s is taken as 2.5 as a representative case for high-modulus CFRP.

Case 3: Choose strengthening scheme (b) with normal-modulus CFRP to show the influence of t_c/t_s as crack propagates (i.e., $2a/W_s$ increases). In this case, W_c/W_s is taken as 0.6 as a representative case for strengthening scheme (b) shown in Figure 7.2(b). E_c/E_s is taken as 1.0 as a representative case for normal CFRP.

Case 4: Choose strengthening scheme (b) with high-modulus CFRP to show the influence of t_c/t_s as crack propagates (i.e., $2a/W_s$ increases). In this case, W_c/W_s is taken as 0.6 as a representative case for strengthening scheme (b). E_c/E_s is taken as 2.5 as a representative case for high CFRP.

Case 5: Choose strengthening scheme (a) with a relatively low t_c/t_s ratio to show the influence of E_c/E_s as crack propagates (i.e., $2a/W_s$ increases). In this case, W_c/W_s can be taken as 1.0 because of strengthening scheme (a). A value of 0.2 is taken for the t_c/t_s ratio representing a relatively thick steel plate strengthened by CFRP.

Case 6: Choose strengthening scheme (a) with a relatively high t_c/t_s ratio to show the influence of E_c/E_s as crack propagates (i.e., $2a/W_s$ increases). In this case, W_c/W_s can be taken as 1.0 because of strengthening scheme (a). A value of 0.5 is taken for the t_c/t_s ratio representing a relatively thin steel plate strengthened by CFRP.

Case 7: Choose strengthening scheme (b) with a relatively low t_c/t_s ratio to show the influence of E_c/E_s as crack propagates (i.e., $2a/W_s$ increases). In this case, W_c/W_s is taken as 0.6 as a representative case for strengthening scheme (b). A value of 0.2 is taken for the t_c/t_s ratio representing a relatively thick steel plate strengthened by CFRP.

Case 8: Choose strengthening scheme (b) with a relatively high t_c/t_s ratio to show the influence of E_c/E_s as crack propagates (i.e., $2a/W_s$ increases). In this case, W_c/W_s is taken as 0.6 as a representative case for strengthening scheme (b). A value of 0.5 is taken for t_c/t_s ratio representing a relatively thin steel plate strengthened by CFRP.

The above examples are plotted in Figure 7.13(i) to (viii). The following observations are made:

- In general, the SIF reduces as t_c/t_s or E_c/E_s increases.
- More reduction in SIF is obtained for double-sided repair than that for single-sided repair.
- More reduction in SIF is obtained for high-modulus CFRP than that for normal-modulus CFRP.
- More reduction in SIF is obtained for relatively thin steel plates than that for thicker plates.
- The SIF reduction ratio is more sensitive to $2a/W_s$ for strengthening scheme (a) than for scheme (b).

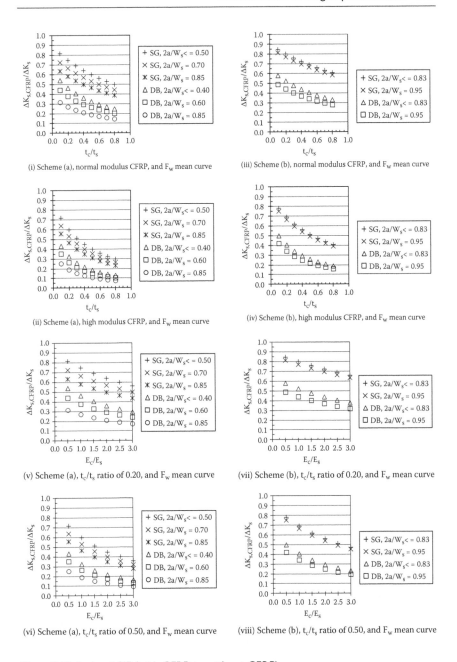

(i) Scheme (a), normal modulus CFRP, and F_w mean curve

(ii) Scheme (a), high modulus CFRP, and F_w mean curve

(iii) Scheme (b), normal modulus CFRP, and F_w mean curve

(iv) Scheme (b), high modulus CFRP, and F_w mean curve

(v) Scheme (a), t_c/t_s ratio of 0.20, and F_w mean curve

(vi) Scheme (a), t_c/t_s ratio of 0.50, and F_w mean curve

(vii) Scheme (b), t_c/t_s ratio of 0.20, and F_w mean curve

(viii) Scheme (b), t_c/t_s ratio of 0.50, and F_w mean curve

Figure 7.13 Ratio of SIF (with CFRP to without CFRP).

7.7 FUTURE WORK

Research is needed to study the effect of fatigue loading on bond strength between steel and ultra-high-modulus (UHM) CFRP plates, and the increase in fatigue life if such UHM CFRP plates are used. Preliminary results (Wu et al. 2012a) showed promising strengthening efficiency. Work is being carried out (Wu et al. 2012c) to develop SIF formulae for CFRP-strengthened steel plates when the CFRP plates do not cover the initial crack region.

Most of the studies described in this chapter were based on idealised cracks in steel members. Investigations are needed to understand the crack propagation in welded specimens where residual stresses and boundary conditions may play an important role. Typical welded connection types may include those identified in plate girder and box girder bridges, steel orthotropic bridges (JSSC 1995), structural systems in undercarriages of motor vehicles and trailers (Xiao and Zhao 2012), truss bridges with cast steel joints (Schumacher and Nussbaumer 2006), and tubular joints in off-shore and mining equipment (Chiew et al. 2004, Pang et al. 2009). There is also a need to study FRP repairing riveted bridges (Bassetti et al. 2000).

Repair could occur at various stages of service life in ageing structures with different damage levels. It is necessary to study the influence of damage level on FRP strengthening efficiency. There is a need to investigate the methods to monitor the fatigue crack propagation in steel structures repaired by CFRP, e.g., through strain gauges on the surface of CFRP (Nakamura et al. 2008) or applying crack propagation gauges on the surface of the steel underneath the CFRP, which can give an indication of where the crack tip is.

REFERENCES

Albrecht, P., and Yamada, K. 1997. Rapid calculation of stress intensity factors. *Journal of the Structural Division*, 103(ST2), 377–389.

Aliabadi, M.H., and Rooke, D.P. 1991. *Numerical fracture mechanics*. Southampton, UK: Computational Mechanics Publications and Kluwer Academic Publishers.

Anderson, T.L. 2005. *Fracture mechanics: Fundamentals and applications*. 3rd ed. London: Taylor & Francis.

ASTM E647. 2001. *Standard test method for measurement of fatigue crack growth rates*. Annual Book of ASTM Standards. West Conshohocken, PA: ASTM.

Baker, A.A., Rose, L.R.F., and Jones, R. 2002. *Advances in the bonded composite repair of metallic aircraft structure*. Amsterdam: Elsevier.

Bassetti, A., Nussbaumer, A., and Hirt, M.A. 2000. Crack repair and fatigue extension of riveted bridge members using composite materials. In *Proceedings of Bridge Engineering Conference (ESE-IABSE-FIB)*, Sharm El Sheikh, Egypt, March 26–30, pp. 227–238.

BEASY. 2006. *Beasy user's guide for revision 9.0.* Southampton, UK: Computational Mechanics BEASY Ltd.

Buyukozturk, O., Gunes, O., and Karaca, E. 2004. Progress on understanding debonding problems in reinforced concrete and steel members strengthened using FRP composites. *Construction and Building Materials*, 18(1), 9–19.

Chen, T., Yu, Q.Q., Gu, X.L., and Zhao, X.L. 2012. Study on fatigue behavior of strengthened non-load carrying cruciform welded joints using carbon fibre sheets. *International Journal of Structural Stability and Dynamics*, 12(1), 179–194.

Chiew, S.P., Lie, S.T., Lee, C.K., and Huang, Z.W. 2004. Fatigue performance of cracked tubular T-joints under combined loads. I. Experimental. *Journal of Structural Engineering*, 130(4), 562–571.

Colombi, P. 2004. On the evaluation of compliance information for common crack growth specimens reinforced by composite patch. *International Journal of Fracture*, 125(1), 73–87.

Colombi, P. 2005. Plasticity induced fatigue crack growth retardation model for steel elements reinforced by composite patch. *Theoretical and Applied Fracture Mechanics*, 43(1), 63–76.

Colombi, P., Bassetti, A., and Nussbaumer, A. 2003. Analysis of cracked steel members reinforced by pre-stress composite patch. *Fatigue and Fracture of Engineering Materials and Structures*, 26(1), 59–66.

Cruse, T.A., and Besuner, P.M. 1975. Residual life prediction for surface cracks in complex structural details. *Journal of Aircraft*, 12(4), 369–375.

Dowrick, G., Cartwright, D.J., and Rooke, D.P. 1980. The effects of repair patches on the stress distribution in a cracked sheet. In *Proceedings of the Second International Conference on Numerical Methods in Fracture Mechanics*, Swansea, UK, July 7–11.

Duong, C.N., and Wang, C.H. 2004. On the characterization of fatigue crack growth in a plate with a single-sided repair. *Journal of Engineering Materials and Technology*, 126(2), 192–198.

Duong, C.N., and Wang, C.H. 2007. *Composite repair—Theory and design*. Oxford: Elsevier.

Elber, W. 1970. Fatigue crack closure under cyclic tension. *Engineering Fracture Mechanics*, 2(1), 37–45.

Fawzia, S. 2007. Bond characteristics between steel and carbon fibre reinforced polymer (CFRP) composites. PhD thesis, Department of Civil Engineering, Monash University, Melbourne, Australia.

Grundy, P. 2004. Fatigue design of steel structures. *Steel Construction*, 38(1), 1–8.

Gu, L., Kasavajhala, A.R.M., and Zhao, S. 2010. Finite element analysis of cracks in aging aircraft structures with bonded composite-patch repairs. *Composites Part B: Engineering*, 42(3), 505–510.

Hosseini-Toudeshky, H., and Mohammadi, B. 2007. A simple method to calculate the crack growth life of adhesively repaired aluminum panels. *Composite Structures*, 79(2), 234–241.

Ingraffea, A.R., Blandford, G.E., and Ligget, J.A. 1983. Automatic modelling of mixed-mode fatigue and quasi-static crack propagation using the boundary element method. In *Proceedings of 14th Symposium on Fracture Mechanics*, ASTM STP 791, ed. Lewis, J.L., and Sines, G., pp. 407–426.

James, L., and Anderson, W. 1969. A simple experimental procedure for stress intensity factor calibration. *Journal of Engineering Fracture Mechanics*, 1(4), 565–568.

Jiao, H., Zhao, X.L., and Mashiri, F.R. 2012. Improving fatigue performance of CFRP strengthened steel beams by applying vacuum pressure in the wet layup of CFRP woven sheets. Presented at Proceedings of the Third Asia-Pacific Conference on FRP in Structures (APFIS2012), Sapporo, Japan, February 2–4.

Jones, S.C., and Civjan, S.A. 2003. Application of fibre reinforced polymer overlays to extend steel fatigue life. *Journal of Composites for Construction*, 7(4), 331–338.

JSSC. 1995. *Fatigue design recommendations for steel structures*. Tokyo: Japan Society of Steel Construction.

Lee, W.Y., and Lee, J.J. 2004. Successive 3D FE analysis technique for characterization of fatigue crack growth behavior in composite-repaired aluminum plate. *Composite Structures*, 66(1–4), 513–520.

Lie, S.T., Lee, C.K., Chiew, S.P., and Shao, Y.B. 2005. Validation of surface crack stress intensity factors of a tubular K-joint. *International Journal of Pressure Vessels and Piping*, 82(8), 610–617.

Liu, H.B. 2008. Fatigue behaviour of CFRP reinforced steel plates. PhD thesis, Department of Civil Engineering, Monash University, Melbourne, Australia.

Liu, H.B., Al-Mahaidi, R., and Zhao, X.L. 2009a. Experimental study of fatigue crack growth behaviour in adhesively reinforced steel structures. *Composite Structures*, 90(1), 12–20.

Liu, H.B., Zhao, X.L., and Al-Mahaidi, R. 2009b. Boundary element analysis of CFRP reinforced steel plates. *Composite Structures*, 91(1), 74–83.

Liu, H.B., Xiao, Z.G., Zhao, X.L., and Al-Mahaidi, R. 2009c. Prediction of fatigue life for CFRP strengthened steel plates. *Thin-Walled Structures*, 47(10), 1069–1077.

Liu, H.B., Zhao, X.L., Al-Mahaidi, R., and Chiew, S.P. 2007. Experimental fatigue crack growth of steel plates with a single-sided composites repair. Presented at Proceedings of 2nd International Maritime—Port Technology and Development Conference, Singapore, September 26–28.

Mashiri, F.R., Zhao, X.L., and Grundy, P. 2000. Crack propagation analysis of welded thin-walled joints using boundary element method. *Computational Mechanics*, 26(2), 157–165.

Mi, Y., and Aliabadi, M.H. 1992. Dual boundary element method for three-dimensional fracture mechanics analysis. *Engineering Analysis with Boundary Elements*, 10(2), 161–171.

Mi, Y., and Aliabadi, M.H. 1994. Three-dimensional crack growth simulation using BEM. *Computers and Structures*, 52(5), 871–878.

Naboulsi, S., and Mall, S. 1996. Characterization of fatigue crack growth in aluminium panels with a bonded composite patch. *Composite Structures*, 37(3–4), 321–334.

Nadauld, J., and Pantelides, C.P. 2007. Rehabilitation of cracked aluminium connections with GFRP composites for fatigue stresses. *Journal of Composites for Construction*, 11(3), 328–335.

Nakamura, H., Jiang, W., Suzuki, H., Maeda, K., and Irube, T. 2009. Experimental study on repair of fatigue cracks at welded web gusset joint using CFRP strips. *Thin-Walled Structures*, 47(10), 1059–1068.

Nakamura, H., Maeda, K., Suzuki, H., and Irube, T. 2008. Monitoring for fatigue crack propagation of steel plate repaired by CFRP strips. In *Proceedings of the 4th International Conference on Bridge Maintenance, Safety and Management*, Seoul, July, pp. 2943–2950.

Pang, N.L., Zhao, X.L., Mashiri, F.R., and Dayawansa, P. 2009. Full-size testing to determine stress concentration factors of dragline tubular joints. *Engineering Structures*, 31(1), 43–56.

Ratwani, M.M. 1979. Analysis of cracked, adhesively bonded laminated structures. *AIAA Journal*, 17(9), 988–994.

Sabelkin, V., Mall, S., Hansen, M.A., Vandawaker, R.M., and Derriso, M. 2007. Investigation into cracked aluminum plate repaired with bonded composite patch. *Composite Structures*, 79(1), 55–66.

Schumacher, A., and Nussbaumer, A. 2006. Experimental study on the fatigue behaviour of welded tubular K-joints for bridges. *Engineering Structures*, 28(5), 745–755.

Seo, D.C., and Lee, J.J. 2002. Fatigue crack growth behavior of cracked aluminum plate repaired with composite patch. *Composite Structures*, 57(1–4), 323–330.

Shen, H., and Hou, C. 2011. SIFs of CCT plate repaired with single-sided composite patch. *Fatigue and Fracture of Engineering Materials and Structures*, 34(9), 728–733.

Suzuki, H. 2004. Experimental study on repair of cracked steel member by CFRP strip and stop hole. Presented at Proceedings of the 11th European Conference on Composite Materials, Rhodes, Greece, May 31–June 3.

Tada, H., Paris, P.C., and Irwin, G.R. 1985. *The stress analysis of cracks handbook*. 2nd ed. St. Louis, MO: Paris Productions.

Täljsten, B., Hansen, C.S., and Schmidt, J.W. 2009. Strengthening of old metallic structures in fatigue with prestressed and non-prestressed CFRP laminates. *Construction and Building Materials*, 23(4), 1665–1677.

Tavakkolizadeh, M., and Saadatmanesh, H. 2003. Fatigue strength of steel girders strengthened with carbon fiber reinforced polymer patch. *Journal of Structural Engineering*, 129(2), 186–196.

Wang, C.H. 2002. Fatigue crack growth analysis of repaired structures. In *Advances in the bonded composite repair of metallic aircraft structures*, ed. Baker, A.A., Rose, L.R.F., and Jones, R., pp. 353–374. Oxford: Elsevier.

Wen, P.H., Aliabadi, M.H., and Young, A. 2002. Boundary element analysis of flat cracked panels with adhesively bonded patches. *Journal of Engineering Fracture Mechanics*, 69(18), 2129–2146.

Wen, P.H, Aliabadi, M.H., and Young, A. 2003. Boundary element analysis of curved cracked panels with adhesively bonded patches. *International Journal for Numerical Methods in Engineering*, 58(1), 43–61.

Wu, C., Zhao, X.L., Al-Mahaidi, R., and Duan, W.H. 2012b. Mode I stress intensity factor (SIF) of centre-cracked tensile (CCT) steel plates. *International Journal of Structural Stability and Dynamics*, in press.

Wu, C., Zhao, X.L., Al-Mahaidi, R., and Duan, W.H. 2012c. Effects of CFRP bond locations on the mode I stress intensity factor of centre-cracked tensile steel plates. *Fatigue and Fracture of Engineering Materials and Structures*, doi:10.1111/j.1460–2695.2012.01708.x

Wu, C., Zhao, X.L, Al-Mahaidi, R., Emdad, M.R., and Duan, W.H. 2012a. Fatigue tests of cracked steel plates strengthened with UHM CFRP plates. *Advances in Structural Engineering—An International Journal*, 15(10), 1801–1815.

Xiao, Z.G., and Zhao, X.L. 2012. CFRP repaired welded thin-walled cross-beam connections subject to in-plane fatigue loading. *International Journal of Structural Stability and Dynamics*, 12(1), 195–211.

Yang, Z.M., Lie, S.T., and Gho, W.M. 2007. Fatigue crack growth analysis of a square hollow section T-joint. *Journal of Constructional Steel Research*, 63(9), 1184–1193.

Ye, H.W., König, C., Ummenhofer, T., Qiang, S.Z., and Plum, R. 2010. Fatigue performance of tension steel plates strengthened with prestressed CFRP laminates. *Journal of Composites for Construction*, 14(5), 609–615.

Young, A., Rooke, D.P., and Cartwright, D.J. 1992. Analysis of patched and stiffened cracked panels using the boundary element method. *International Journal of Solids and Structures*, 29(17), 2201–2216.

Zheng, Y., Ye, L.P., Lu, X.Z., and Yue, Q.R. 2006. Experimental study on fatigue behavior of tensile steel plates strengthened with CFRP plates. In *Proceedings of Third International Conference on FRP in Composites in Civil Engineering (CICE 2006)*, Miami, FL, December 13–15, pp. 733–736.

Index